£75p

COMMON GROUND

HOW CHANGES IN THE COMMON AGRICULTURAL POLICY AFFECT THE THIRD WORLD POOR

ADRIAN MOYES

First published in January 1987
Reprinted in March 1988

British Library Cataloguing in Publication Data

Moyes, Adrian
Common Ground: how changes in the Common Agricultural Policy affect the Third World poor.
1. Developing Countries. Agricultural industries. Effects of European Economic Community.
I. Titles II. Oxfam
338.1'8'09172

ISBN 0-85598-078-8 Pbk

Published by Oxfam, 274 Banbury Road, Oxford OX2 7DZ, UK.
Typeset by Typo Graphics, Oxford
Printed by Oxfam Print Unit

CONTENTS

INTRODUCTION

In the rich world the problem of food is one of plenty — in the Third World there is not enough. But perhaps things are changing — or at least changeable. In the Common Market — or European Community (EC) — the cost of the food surpluses and their effects on world trade have led to pressures so strong that change of some kind seems inevitable. And Bob Geldof has voiced the concern of millions that for the Third World such shortages of food should never happen again.

It is easy to believe that the surpluses in the EC can be used to relieve the Third World's poverty — or that EC policies have in some degree caused it. (It has been estimated that about 36 million acres in the Third World is used to produce the EC's animal feeds.)[1] Both beliefs are based on only partial truths; the CAP has contributed to some of the poverty, but most of it has other causes, including the policies of other rich and poor countries, and fluctuating exchange rates. And while surpluses can be vital to feed the starving (as the Send-a-Tonne to Africa scheme showed) the effects of the CAP on the poor go far beyond food aid.

For the EC is not self-contained; it is the largest **importer** of agricultural products in the world, and 60% of these imports come from the Third World. Some of them cannot be grown in Europe — tea, coffee, rubber, for instance. But others can — animal feeds, tobacco, cotton and sugar. The EC is also the world's second largest **exporter** of agricultural products. Many of these exports are also grown and consumed in the Third World — sugar, cereals, dairy products, meat. So the EC is both a source of supply and a competitor.

What is grown in the EC therefore, has effects on the crops that many Third World farmers grow in their fields, on the prices that the poor of Third World cities pay for their food, and on the foreign exchange available to Third World governments — for agriculture, for development, for luxuries, or for arms.

Mounting costs and pressures from other trading countries make changes to the CAP almost certain. Equally certainly the changes will affect poor people in the Third World, some of them adversely. It is to highlight these effects, and to suggest what can best be done about them, that the **Enquiry on the Effect of Changes in the Common Agricultural Policy on the Third World** has been set up. It has taken the form of a panel of farmers, politicians, academics and trades unionists which held a series of hearings in agricultural areas of Britain in the summer of 1986. At the hearings, expert witnesses gave evidence — and others who did not attend the hearings also submitted written evidence. A list of the Panel members is given on page 5, summaries of all the evidence are given in the Appendix.

Although the EC is only one of the major factors in world agricultural trade, the Panel concentrated on it — partly because it is so big, especially in relation to the Third World (see Section 1), partly because it is so likely to change, and partly because it is the policy which farmers, politicians and others in Britain are most interested in, and are most likely to be able to influence.

The first section of the report summarises the main features of the CAP. There follow separate sections on **animal feeds, wheat, food aid**, and **sugar**. A final section brings together the Panel's conclusions. A series of five Case-boxes are used to portray the everyday lives of a few of the many people in the Third World who are affected by the CAP. The people in the Case-boxes are real, but their names have been changed.

* * *

The political leaders of Europe have wisely realised that there is more to agricultural policy than food; there are other interests to be considered besides those of farmers. Rural communities deserve protection, farm workers need improved conditions, the environment and the food security of nations is at stake. In this report we have described another interest — that of the Third World. It is one that has been largely ignored to date, but it affects in one way or another perhaps as many people as live in the EC.

The Panel believe that almost any of the changes in the CAP that are on the horizon will bring harm to some poor people in the Third World. To argue for no change is clearly unrealistic — and undesirable. Europe does not possess the optimal agricultural policy at present, and there are legitimate pressures for change. No new CAP will give **primary** consideration to the hungry in the Third World, but the Panel is convinced that with forethought, care and political will it is possible to minimise the harm and to compensate for damage done to the poor.

The Panel has considered what such a strategy might consist of. This report describes its view of a new approach to agricultural policy. It is an approach that is pragmatic, that balances the requirements of the people of the EC with the needs of the Third World's poor, and that could be supported by farmers, politicians and the public of the EC alike.

Farmers in the EC do not want to contribute to other people's misery. On the contrary, the challenge of ensuring that no one in the world — adult or child — goes hungry to bed is the ultimate purpose of farming, and one that farmers everywhere have a part in. This is the shared interest of farmers and consumers in the North and in the South. The hope of the Enquiry is to build on this common ground.

SUMMARY OF FINDINGS

1 The EC's large-scale imports and, more recently, exports, make it a major force in world agricultural trade. Any change in the CAP therefore, will affect poor people in the Third World — many of them unfavourably.

2 Besides looking after its own people, the EC has as responsibility to treat its trading partners fairly — especially those who have little power and a lot to lose.

3 The EC should therefore establish machinery for identifying who in the Third World is likely to be hurt by any proposed change, and should present such information to EC decision-makers.

4 The EC should accept the principle of making good the harm it causes to people in the Third World through any change in the CAP.

5 It is in the interests of the rich countries that the poor should get richer; as incomes increase, so does profitable trade.

6 The EC's surpluses are in general damaging to the Third World; they should be reduced, eventually to zero. The method of reduction is important; some methods would reduce Third World markets or lower the prices of Third World exports.

7 The CAP harms Third World countries by increasing world price instability. This could be reduced if the CAP were made more responsive to world markets.

8 Wherever possible the EC should offer the market to those who need it. It should not grow (expensively) for itself products which can be grown (more cheaply) in the Third World and provide an income for people who may have few alternatives. The EC should maintain its imports of sugar from the Third World at near CAP prices.

9 The EC should help Third World countries get the maximum value from what they produce by cutting the import taxes (tariffs) it currently charges on processed agricultural products. Such cuts would have effects on employment and profits in the EC, and they would have to be made with care.

10 In the forthcoming negotiations on agricultural trade at GATT (the General Agreement on Tariffs and Trade) the rich countries should listen to the voices of the Third World.

11 The EC should open membership of the Lomé Convention to any of the ten 'least developed countries' not currently members who wish to join.

12 Food aid should not be used to justify the EC's surpluses. Its use in emergencies should be increased. Its organisation should be improved.

MEMBERS OF THE PANEL

Jack Boddy
National Secretary, Agricultural and Allied Workers National Trade Group.

Margaret Daly
Conservative MEP for Somerset, member of the European Parliament's Development and Cooperation Committee.

Tony Leeks
Visiting Fellow at the Institute of Development Studies, Sussex University, former Director, Trade and Commodities Division, FAO.

Richard Livsey
Member of Parliament for Brecon and Radnor, Liberal Spokesman on Agriculture.

Adrian Moyes, Secretary
Development Secretary, Oxfam, formerly Field Director, Tanzania and former head of Oxfam's Public Affairs Unit.

Derek Pearce
Farmer and business consultant, Master of the Worshipful Company of Farmers.

John Quicke
Farmer, cheese manufacturer, Countryside Commissioner, past President Country Landowner's Association.

David Richardson, Chair
Farmer, Presenter of 'Farming Diary' on Anglia TV and Columnist for 'Big Farm Weekly'.

Chris Stevens
Research Fellow at the Centre for European Policy Studies in Brussels and at the Institute of Development Studies at Sussex University, and Research Officer at the Overseas Development Institute, London.

John Tomlinson
Labour MEP for Birmingham West, Member of Euro-Parliament Budgets Committee.

1 THE COMMON AGRICULTURAL POLICY

"In the United States, the government pays farmers not to grow grain; in the European Community, farmers are paid high prices even if they produce excessive amounts. In Japan rice farmers receive three times the world price for their crop; they grow so much that some of it has to be sold as animal feed — at half the world price. In 1985, farmers in the EC received 18 US cents a pound for sugar that was then sold on the world markets for 5 cents a pound; at the same time, the EC imported sugar at 18 cents a pound. Milk prices are kept high in nearly every industrial country, and surpluses are the result; Canadian farmers will pay up to eight times the price of a cow for the right to sell that cow's milk at the government's support price. The United States subsidises irrigation and land clearing projects and then pays farmers not to use the land for growing crops."

World Bank, World Development Report, 1986[2]

Many countries, developed and developing, protect and subsidise their agriculture. The EC is not unique — and it is not the only cause of the effects that protectionist policies have on the Third World. But it is one of the most important causes, because it is one of the world's major trading blocks — particularly with the Third World.

* * *

"The CAP was and remains the 'marriage contract' of the EC" as the EC Commission puts it.[3] There are two main reasons for this:

— the agricultural policies of member countries differed markedly, resulting in different prices for food. Since the price of food affects wages, it also affects industrial competitiveness between member countries.

— France, Italy and the Netherlands would only agree to let in German industrialised goods in return for Germany taking their agricultural exports.

Even so, the CAP had a stormy genesis; it took seven years to get agreement on the key measure of cereal prices, and then (1964) only after the entire Commission had threatened to resign.[4] Its basic aims are to

— **improve food security** — really a form of insurance against war, weather or pests. This aim leads to the objective of greater **self-sufficiency**, achieved through higher agricultural productivity.

— maintain **jobs on the land**, preferably on family farms,

— ensure a **"fair standard of living for the agricultural community"**

— **stabilise markets**

— **keep consumer food prices stable and reasonable**

The EC contains 321 million people and 10.7 million farmers. But the agricultural community that the CAP caters for is rather different from that in some of the Member states, such as Britain. In 1984 51% of the holdings in the EC were less than 50 acres, indeed 30% were less than 12 acres. The addition of Spain and Portugal has increased the number of small farmers.[5]

All developed countries support their agriculture; the EC's arrangements are described in Box 1.A.

Pressures for change

There are pressures for change from both inside and outside the EC.

On the inside, the main worry is the **cost**. The budget for the CAP for 1986 is £14,638 million — not only a big sum in itself, but nearly two-thirds of the total EC budget.[6]

The indirect cost may well be higher. Because of the CAP, agriculture in the EC is more profitable than it would be without it — so investment is diverted into land, machinery, research etc, which would otherwise go into manufacturing industry. As usual, there are different estimates of the scale of the effects of this; one study suggests that since 1973 the CAP has reduced the EC's manufactured output by 1.5% and its manufactured exports by 4%; has increased its manufactured imports by 5% — and resulted in an extra 1½ million unemployed.[7]

People are also worried about what the money is spent on. Some goes to subsidising exports — £406 million on wheat throughout the EC in 1985, £1,401 million on dairy products. Some goes to storing the surplus; the 'mountains' and 'lakes' cost £2,065 million in 1985.[8] Also, in the EC as in most developed countries, the main beneficiaries of agricultural protection are land-owners and quota-holders; poor consumers bear a disproportionate share of the cost because they spend a large share of their income on food. The CAP is an inefficient way of transferring money between various sections of society; the World Bank estimates that for every £1 that farmers gain, taxpayers and consumers pay £1.50.[9] Nor do all farm-workers gain — indeed many are particularly low-paid.[10]

A third worry, particularly strong in Britain and the Netherlands, concerns the effect of intensive farming on the countryside. Many are concerned that as fields are enlarged and drained, hedges, woods, heaths and ponds are lost. Pesticides affect humans, wild animals, insects and plants. These costs could perhaps be accepted if that were the only way to produce enough food. But in the face of the surpluses they are seen by many as increasingly unacceptable. Some of the suggested changes to the CAP, such as lower prices for cereals might lead to greater use of pesticides and fertilisers as farmers attempted to grow more and maintain their incomes.

All these concerns have been well publicised. Some farmers report that nowadays they are made to feel that growing food is something disreputable, to be ashamed of.[11]

At the same time farmers themselves are worried by falling incomes and investment, high borrowings, and general uncertainty about the future of the CAP.

Another pressure for change comes from the two new members of the EC, Spain and Portugal — both at the poor end of the EC spectrum, and likely to require some of the resources of the richer members.

On the outside, the pressure for change comes from other agricultural exporters, most importantly from the USA, which sees subsidised exports from the EC both competing for markets and lowering prices. Since the 1950s, the USA has built up its food exports with strong government support (farm support in the USA is probably as high as in the EC — see Box 3.B, p 38 for details). It has also used food aid to generate subsequent food sales. It now exports some 45% of its harvested acreage (and hopes to increase it to 50% by the year 2000).[12] A recent US law (the Food Security Act, 1985) gives teeth to the policy of subsidised grain sales and seems specifically designed to undercut EC sales of surpluses.

Not surprisingly the USA is concerned about the effect of the CAP on its export prospects. In Spring/85 it announced a £1,300 million plan to subsidise US exports to give them the edge over those subsidised by the EC. But obviously the USA would prefer negotiated changes in the CAP to such expensive measures.

The EC's trade with the Third World

The EC is the largest importer of agricultural products in the world — £23,000 million in 1984, of which 60% came from the Third World. After the USA it is the second largest exporter in the world, with agricultural exports of £15,000 million in 1983. The value of this trade is increasing rapidly — imports by nearly four times between 1965 and 1980, and exports by about eight times.

Agricultural products can usefully be considered in two categories — those products which can be grown in the EC, and those which cannot (such as coffee, tea and cocoa). The CAP affects the first category more strongly, because what farmers in the EC grow may compete with Third World farmers, and because EC farmers could grow for themselves what they currently import from the Third World.

Case-box 1 describes the life of some of the people producing these imports — a family of Brazilian cotton-workers. Charts 1 and 2 give a picture of the EC's agricultural trade. They cover external trade only — not trade between EC members.[13]

Imports

1 The main imports

Over half the EC's total agricultural imports come from the Third World — and the EC takes about a third of their exports.

Coffee is (1983) the largest agricultural import — nearly four times the value of **cocoa**, and nearly nine times that of **tea** (Chart 2).

The EC imports a lot of **animal feeds** because they are cheaper than domestically produced grain.

Most of the imports come in unprocessed form because tariffs (i.e. import taxes) penalise processed goods. This means that Third World countries are not able to get the fullest value from their products. Table 1.1 gives the tariffs for soybeans.

Table 1.1

Tariff barriers on soybeans and soybean products

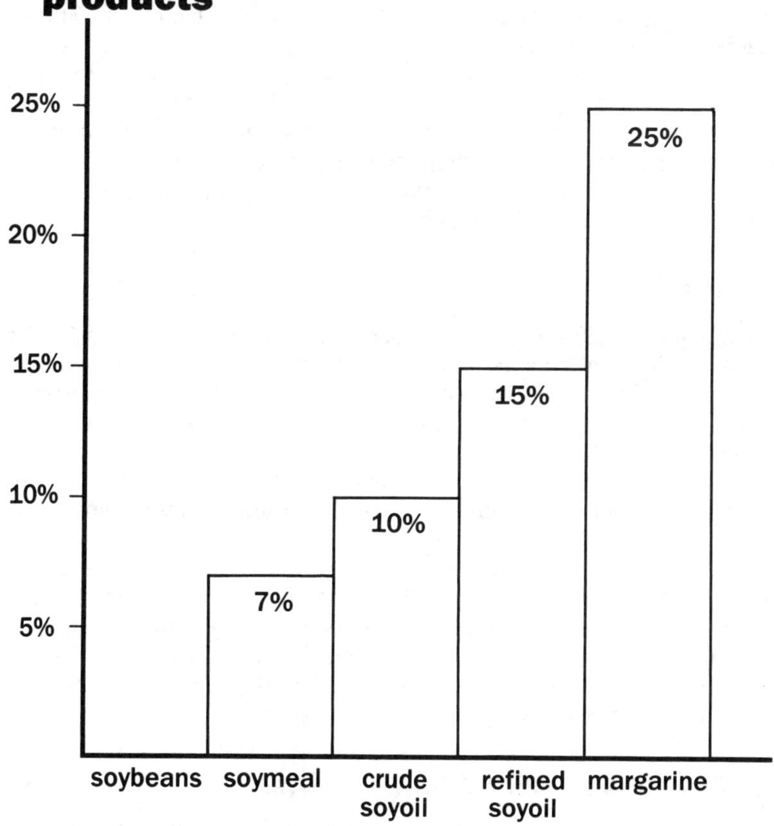

Source: Evidence to the Enquiry, Geoffrey Bastin.

These (and many similar) taxes on processed agricultural products dramatically reduce the potential income of Third World countries and, as the World Bank puts it, *"severely disrupt the process of development."*

2 The main suppliers
Brazil is (1984) far and away the largest Third World supplier; its trade is over three times the value of the next two largest, **Argentina** and the **Ivory Coast** (Chart 1). To many people in Britain, used to thinking of the Third World in terms of the Commonwealth, the scarcity of Commonwealth members amongst the biggest suppliers is surprising.

The importance of the EC's trade to Third World countries can be seen by expressing it in terms of the exporting country's population. Here the **Ivory Coast** leads, with an astonishing £91 worth of exports to the EC per head of population. The Philippines (£57) and Cameroun (£37) are also high.

3 The importers
Germany is (1983) the largest importer (3,600 million, or £59 per head of population), followed by **France** (£2,700 million — £50 per head) and **Britain** (£2,600 — £46 per head).

These figures show the point of entry into the EC, not the ultimate destination. Some imports — those of the Netherlands, with its huge Euro-port at Rotterdam, in particular — are re-exported to other member countries.

France imports the highest proportion from the Third World (57%), followed by the Netherlands (54%) and Germany (51%). Britain (45%) is a bit below the average for the whole of the EC (48%).

Exports

Nearly half of the EC's agricultural exports go to the Third World, making 28% of their agricultural imports. **Dairy products** are the most valuable (£1,473 million in 1983), followed by **cereals** (£1,258 million). Note that these figures are for 1983, before the milk quotas were imposed; dairy exports have fallen as a result of the quotas.

A major effect of the EC's exports is their influence on world prices. This effect does not depend on the destination of the export — it is obtained simply by putting the product on the market.

Britain is the third biggest agricultural exporter to the Third World (£1,099 million in 1983), after France (1,962 million) and the Netherlands. These figures show the exporting country, not the country of origin, so the Netherlands figure in particular includes some re-exports. Benelux (Belgium, Netherlands and Luxembourg) and Ireland export the highest proportion of their total agricultural exports to the Third World (66% and 58%).

Concessions to Third World countries

The CAP allows a number of concessions to certain Third World countries, for example the Mediterranean, and a group of former colonial countries under the Lomé Convention. (see Box 1B)

Net trade

The success of the CAP in producing more food than consumers in the EC need has resulted in sizeable exports and quantities of food aid. The publicity attached to this easily gives rise to the impression that the EC is a net supplier of food to the Third World. In fact it is the Third World which is a net supplier of agricultural products to the EC — it exports about £7,000 million a year more to the EC than it imports, even without counting wood, rubber or cotton; to Britain it exports £1,500 million more than it imports.

The effects of the CAP

The CAP affects the poor in the Third World in three main ways — through its exports to them and to the world market, through its imports from them, and through its effect on price stability. Just how it affects any given country depends on the mix of agricultural products produced, imported and exported. How it affects the poor depends also on local government policies and on how the crops are produced — on plantations or small-holdings, for example.

The Enquiry focused on four of the major areas where the CAP affects the Third World — the import of animal feeds, the export of wheat, the import and export of sugar, and food aid. They are dealt with in turn in the following sections.

Box 1.A

HOW THE EC PROTECTS ITS AGRICULTURE

The system the EC has devised and refined over the years in order to protect its agriculture is a complex one. Basically the Common Agricultural Policy gives farmers a guaranteed price for their produce, controls imports which might compete, subsidises exports, and limits or encourages production. The CAP covers most products that can be grown in the EC — except potatoes.

Here is a very simplified explanation.

Prices

The Commission sets a **'Target Price'** (reviewed annually) — which is the price it wants producers to obtain. Any crops which cannot be sold on the EC market are bought 'into intervention' at the **'Intervention Price'**, which is usually only a little lower than the Target Price. Unless limited by Quota (see below), farmers may produce as much as they like with a guaranteed sale. This has led to the surpluses.

Box 1.A cont.

Imports

Since EC prices are usually higher than world prices, it is necessary to prevent cheaper imports from competing with EC produce. This is achieved by **import taxes** (also known as levies, duties or tariffs) and **quotas.**

Import taxes

The most important import tax is a system of **variable levies** which ensure that however cheaply outsiders sell, they are taxed up to the EC price. This system is used for cereals, dairy products, sugar, rice and olive oil.

There are also fixed duties and tariffs which are based on the value of the import. Because they are based on value, they bear heavily on processed agricultural products and thus discourage Third World countries from processing their own products before export. Because they are fixed, EC consumers pay less if outside prices fall. Sometimes duties are combined with a variable levy (as for pork and poultry — which are classified as 'processed cereals'), or with a 'Reference Price', which sets a minimum import price and is used for wine and some fruit and vegetables.

Quotas

In a few, but important, cases the EC sets absolute limits to the amount of a product that may be imported. There are quotas for sugar, beef and cassava. Sometimes, as with sugar, imports within the quota may be at a special price.

Exports

The EC's policy of providing farmers with a guaranteed price no matter how much they produce, has led to surplus production. In order to encourage sales on the world market, the EC has a system of '**Export Restitutions**' (or refunds); the trader sells products competitively on the world market and the EC pays the difference between the EC price and the world price.

Production

In a few cases the EC limits or encourages production. Quotas are used to control production; absolute quotas in the case of milk, price quotas in the price of sugar, where production over the quota is allowed, but price support is not paid.

Other types of control such as limiting the input of fertilisers or the acreage of land (a key part of the US system) are under discussion.

The EC also encourages production of some products, by the use of '**Production Aids**', through which the farmer is paid the difference between the support price and the world market price. This gives a good

Box 1.A cont.

income to the farmer, while keeping consumer prices low. It is used for tobacco (the EC grows 40% of its own tobacco), cotton (at a 1986 cost of £132 million), olive oil and durum wheat.

Box 1.B

THE LOMÉ CONVENTION

The Lomé Convention is a trade and aid agreement between the EC and 66 Third World Countries. They are all in Africa, the Caribbean or the Pacific (there are no Asian ones) and are thus often called ACP countries. They are mostly (apart from Nigeria and Kenya) small and poor, and all have colonial connections with Europe. Their combined population of 368 million is a little less than a tenth of the total Third World population (about 4,000 million).

Nearly all are very dependent on the EC as a market for their exports — around 50% in most cases, up to 80% for some of the smaller ones such as Barbados and Mauritius.

There are three main parts to the Convention:

1 Concessions on imports from ACP countries

ACP countries are given duty-free access for some of their products and more favourable treatment than other countries for others. In most cases these concessions are not very valuable (either the ACP countries do not export much of the products involved, or the tariffs are already low) but some, especially those for sugar (see Box 5.C page 60) and beef are extremely valuable, providing major sources of income for the producing countries.

2 The European Development Fund — the EC's aid programme.

3 STABEX

STABEX is an export earnings stabilisation scheme designed to compensate ACP countries when the price of their major exports falls rapidly. The EC spends over £100 million/year on STABEX. The main products supported are cotton, sisal, coffee, cocoa and groundnuts — none of them covered by the CAP, except cotton. The main beneficiaries in 1983 were:

	Million £
Senegal	51
Sudan	40
Ivory Coast	22
Mauritania	20
Tanzania	15

Sources: 'LOME III', The Courier. No. 89, Jan–Feb/85. Brussels
World Development Report, 1986. IBRD, Washington.

Case-box 1

THE COTTON WORKERS — BRAZIL

Brazil is the fifth largest producer of cotton in the world. The EC currently subsidises its own domestic production of 430,000 tonnes. Any change in this level of subsidy — up or down — could affect cotton-workers in Brazil

Cotton-workers in Brazil have a rough time; they have to contend with low wages, seasonal work and pesticide poisoning. But the income from cotton at least enables them to get by.

* * *

Dona Lia and Senhor Ze-Maria (both 50) live with the four youngest of their seven children aged between 10 and 19. The 18-year-old works in the field and contributes money, but keeps two days pay for his own use. The 13-year-old (disabled) son is already helping in the cotton picking season. Their 28-year-old daughter works in a hotel in the town earning 400 cruzados (CZS), (£20) a month. She usually gives her money to the family, but this month she has had to spend it on her teeth.

The family lives in an adobe brick house about 10′ × 25′ which they built themselves. They have wired it up for electricity and are waiting for it to be connected — they are one of the few areas in the neighbourhood that does not have electricity. Water comes from a covered well; the municipality is slowly putting in piped water but the supply is very irregular. The only sanitation is a pit lavatory.

The house has a 'sitting room' and a kitchen. The walls of the sitting room are adorned with the pictures of saints and spiritist leaders.

Life is hard. A day's work begins at 4.30 am, *"It makes you want to cry"* says Dona Lia *"as you see women bustling along with their little ones bundled up against the cold before work"*. Women prefer to leave their children with neighbours, friends or private child-minders *"guardadeiras"* (75p a day — about a third of the average labourer's daily pay). Or else they leave the kids unattended — *"There aren't too many criminals around, so it's OK"*. There is a free creche in town run by the Rotary club — it provides a bus that collects the children from the street corners in the mornings — but there are only 90 places.

At 5 am the workers wait for the contractors on the street corners — huddled against the cold in what clothes they can afford. The drive in the backs of lorries lasts for up to an hour, or maybe two — along dirt roads. Accidents are common, *"The drivers are crazy and they're always looking at the curves in the women rather than the curves in the road. . . ."*

Around 11 am they take a quick break to eat. Dona Lia complains bitterly that she can't stomach cold food any more. *"I just wish I could retire — I can't bear cold beans any more — they just don't go down"*.

During the week they work until 5 pm or so, not getting home until 6

Photo: Frances Rubin

Young workers returning home.

Chart 1

EC Agricultural Imports – 1984

EC Agricultural Imports from

a) **World**

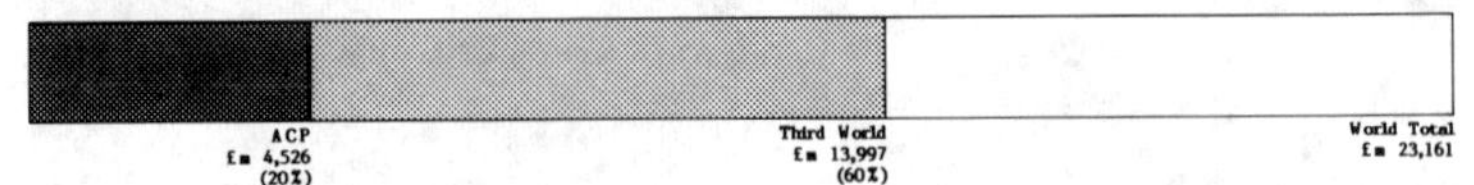

b) **The 14 largest Third World exporting countries**

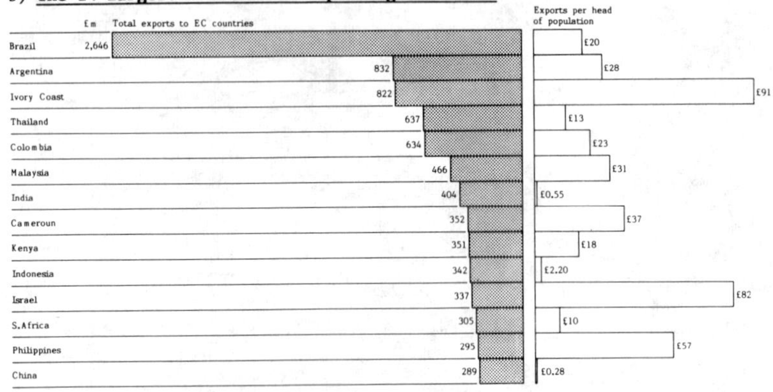

source: Euro-stat 5 – 1985, Table 12: EC Trade by Commodity Classes and main countries.

notes:
i) 'Agricultural Imports' are defined as SITC Classes 0, 1 and 4. This excludes natural rubber, wood and cotton.
ii) 'Third World' is defined as Class 2 countries plus S.Africa and China.
iii) 'Imports' are defined as products entering the EC, and do not include imports from other EC countries (i.e. Intra-EC trade).

Chart 2

Main EC Agricultural Imports from Third World countries – 1983

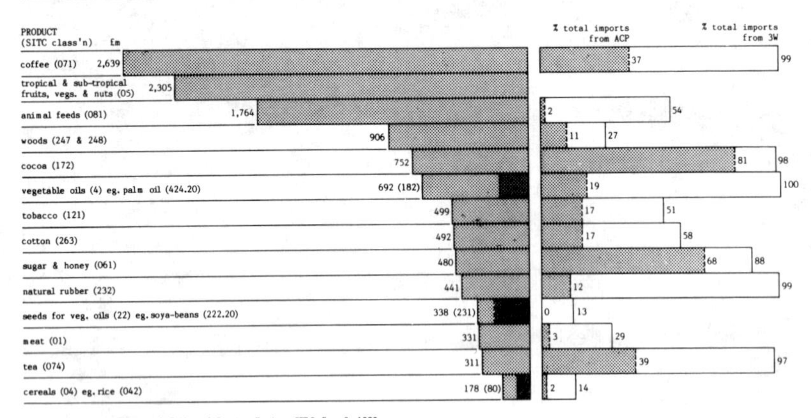

source: Euro-stat Analytical Tables of Foreign Trade – SITC Rev 2, 1983 Import II, 0-4, Table 1.

note: 'Imports' are defined as products entering the EC, and do not include imports from other EC countries (i.e. Intra-EC trade).

The cotton workers cont.

or 7 pm depending on transport. Saturdays are better as they finish a bit earlier. Sunday is the only day off; that's when Dona Lia catches up on the washing etc. *"Holidays? That's only when we don't have work".*

The cotton harvest is paid on a piece rate basis. An ordinary picker gets about the average daily labourer's wage of CZS 50 (£2.50): an experienced picker can make three times that.

Each day's work is paid in vouchers. These can be exchanged for food at local corner shops or supermarkets (and where workers tend to run up debts), or else on Sundays for cash from the contractors. The importance of the cotton harvest is that it's a way of zeroing debt. *"If we don't work hard in the cotton season we might as well give up!"* Dona Lia thinks it's the best work because you can go at your own pace (but you have to be *"OK in the joints"*)

The most common problem, they say, is *"stroke"*. Men, they say are finished by the early 30s. One of the problems is the poison in the fields — *"Sometimes we're working when they come along and spray".* Defoliants are used to speed up the process of bud-opening. *"If someone gets sick in the fields — too bad — he'll have to wait until the lorry comes to take us home".* It's not unheard of for people to die before reaching the town.

Their household income fluctuates; there are times when nothing is coming in. It is probably on the generous side to say that they are getting around 2–3 minimum salaries a month — CZS 2,400 or £120.00 — between the six of them. They spend almost all of this on enough food *"to eat badly".*

There is hardly anything left for clothes, transport, health, entertainment.

Dona Lia's dream? Well she would like to go to the sanctuary of the patron saint of Brazil, "Our Lady of Aparecida", but even more than that she'd like to go and see "Bom Jesus dos Milagres", in the city of Sao Paulo. The days when she hasn't got work she goes to her daughter's house to watch a TV programme at 7 am, showing a "Padre" in Sao Paulo who performs miracles — she wants to go and pay a visit to his sanctuary in the capital because she has a kidney problem. . .

Source: Frances Rubin, Oxfam Recife

2 ANIMAL FEEDS

The Evidence

The EC is the world's largest buyer of animal feeds — nearly 40 million tonnes from all sources in 1984. About 60% comes from the Third World (see Box 2.A). It has been estimated that some 36 million acres in the Third World are devoted to producing the animal feeds that are sent to the EC.[14] Case-box 2 describes the life of soyabean growers in Brazil.

Animal feeds are also one of the largest categories of agricultural import into the EC. Chart 2 (page 16) shows that they come third after coffee (which cannot be grown in the EC) and the catch-all category of Tropical and Sub-tropical Fruits, Vegetables and Nuts. Box 2.A shows how much of each of the main types of animal feeds is imported, and Box 2.B describes them and lists the main places they come from.

The EC's imports of animal feeds are not only large — they are also likely to change. Since the changes are likely to reduce imports, rather than increase them, the effects in the Third World will be adverse, affecting significant numbers of people.

Animal production on the scale now practised in the EC is entirely dependent on animal feeds, whether to supplement grazing or to replace it. Box 2.C shows which type of animals consume the animal feeds.

When the CAP was negotiated in the early '60s the possibility of growing more than a small proportion of the EC's animal feeds requirements was remote. So, largely to placate US opposition to the CAP, the EC agreed not to tax (or not much) imports of oil-seeds/meals/cakes or 'cereal substitutes', such as cassava or fruit residues. These low levels of protection were written into the GATT (General Agreement on Tariffs and Trade — the main international trade agreement — see Box 6.A, page 70). They are now, therefore, very difficult to change — and the imported products are much cheaper than anything comparable produced under the CAP. Table 2.1 gives the comparative costs of vegetable oil production in five areas; production in the EC is nearly 5 times the cost of that in Indonesia. The Table shows oil-seed costs; oil-cake and meal costs are broadly similar.

The consequence is that the EC animal feed market is not insulated from the world market, so imported animal feeds are available to farmers at much less than the cost of internally produced grains or oilseeds.

Not surprisingly, EC imports of animal feeds have grown rapidly — by four times between 1970 and 1982. Although these cheaper imports suit EC farmers who use them extensively to raise livestock, they do not suit the Commission. Because livestock farmers buy imported cereal substitutes rather than EC-grown cereals, the Commission is landed with the problem of storing surplus cereals or selling them (at a subsidy) on the world market. Nor do they suit all farmers equally; those in France and South Germany, for instance, use

Table 2.1

Costs of vegetable oil production in selected countries — 1985

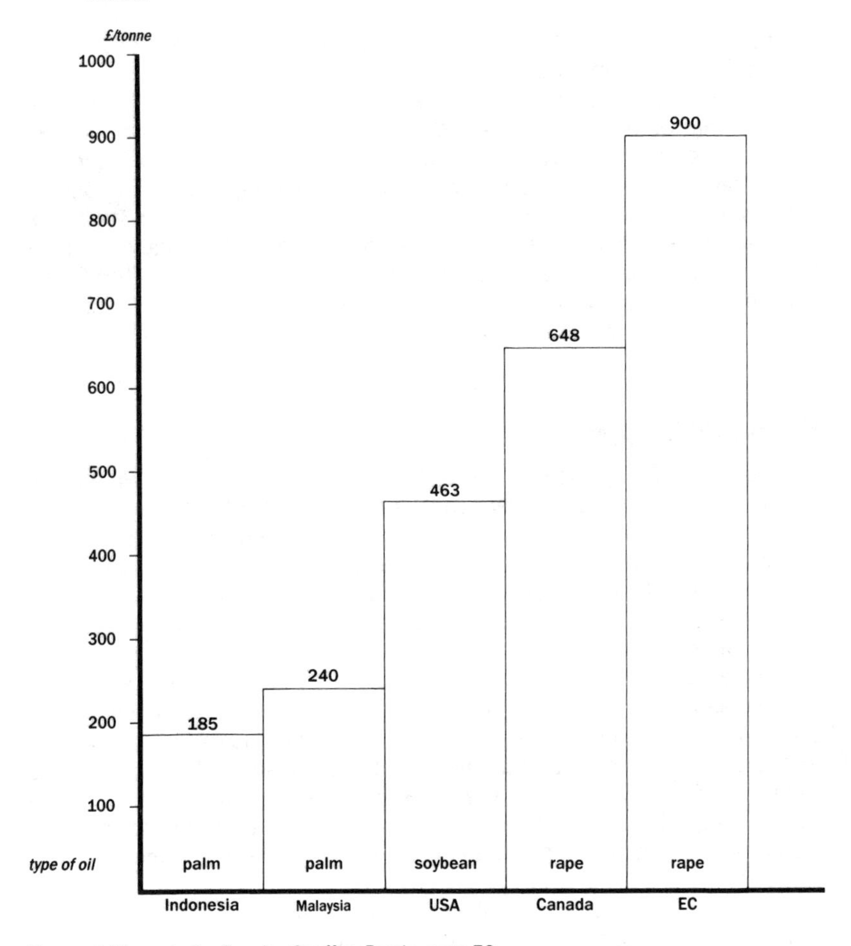

Source: Evidence to the Enquiry, Geoffrey Bastin, page 76.

them much less than those in the Netherlands and North Germany.

The original reason why the Commission was worried about imports of animal feeds was a fear of dependency. In 1973 the USA stopped soyabean exports. The Peruvian fish-meal supply failed, and there was a bad soyabean crop in the USA. Since at that stage the USA supplied over three-quarters of the EC's soyabeans, and soyabeans comprise some two-thirds of all animal feeds (including EC-produced) used in the EC, this aroused serious anxięty. Although the embargo only lasted three weeks, it gave a major boost to the EC's policy of increased self-sufficiency. Since then other factors have entrenched the policy, which now includes a major research programme on rapeseed, field beans and other possible alternatives to imported animal feeds, and a subsidy (currently £170/tonne) on rapeseed.

The EC spent over £700 million on encouraging the domestic production of oilseeds in 1985, some 6% of the total CAP budget.

While the EC is worried about dependency on imports, many Third World countries are worried about over-dependency on particular export crops. The most striking example of this is Thailand — which took advantage of the low EC protection on cereal substitutes to develop a quarter-million-acre cassava industry on which some three million people came to depend — see Box 2.D. In an attempt to limit imports, the EC has negotiated a 'Voluntary Restraint Agreement' which costs Thailand £100 million/year in lost exports. The EC has, however, agreed in aid programme as compensation — though only of £32 million over 5 years (and perhaps not all of that truly 'additional' aid).

What may happen

1 Imports of both vegetable protein and cereal substitutes may fall. Production of both rapeseed and sunflower seeds is rising in the EC. Research on field beans and other alternatives is under way. At the same time there is pressure to reduce production of cereals (to lower the scale and cost of the surpluses) — perhaps by switching to alternative crops such as rapeseed. The likelihood is therefore that there will be a steadily increasing supply of EC-grown vegetable protein for animal feeds.

On the other side, the demand for animal feeds may fall; livestock numbers may decrease. The introduction of milk quotas has already reduced the number of dairy cattle in the EC, attempts to reduce the beef surplus may be successful — and demand from consumers may fall; the trend seems to be to eat less meat. Any of these would reduce livestock numbers, especially of cattle. At the same time genetic improvements may lead to the more efficient use of animal feeds — so less may be needed to produce a given quantity of meat or dairy products.

The supply of EC-grown animal feeds may rise therefore — and the total EC demand for animals, and thus for animal feeds may fall.

2 Cereal prices in the EC may fall; if they do, so will the price of animal feeds. Changes in the CAP designed to reduce the grain surplus may result in lower prices. If they do, EC farmers would reduce their use of imported animal feeds — both vegetable protein and cereal substitutes. But grain prices would have to fall a long way before serious inroads were made on cassava imports.

3 Even if the price of animal feeds falls, Third World countries will continue to export. Many animal feeds are by-products — principally of vegetable oils or cotton. So if farmers are producing the main product, it still pays them to sell the by-product even at a very low price — indeed, in order to continue with the

main product, they have to dispose of the by-product somehow. Third World countries, therefore, keen to get as much foreign exchange as possible, will probably continue to export, even if the price falls considerably.

This does not apply to cassava (a human food in its own right), soya (which is produced primarily for animal feed — usually more than 60% of its value), or fishmeal (which is produced entirely for animal feed).

The effect on the Third World

As usual, some countries and people would gain (from lower prices for the animal feeds they import) and some would lose (from lost markets and lower prices for the animal feeds they export). Table 2.2 shows the main Third World gainers and losers if EC imports declined.

Table 2.2

Main gainers and losers if EC imports of Animal Feeds declined

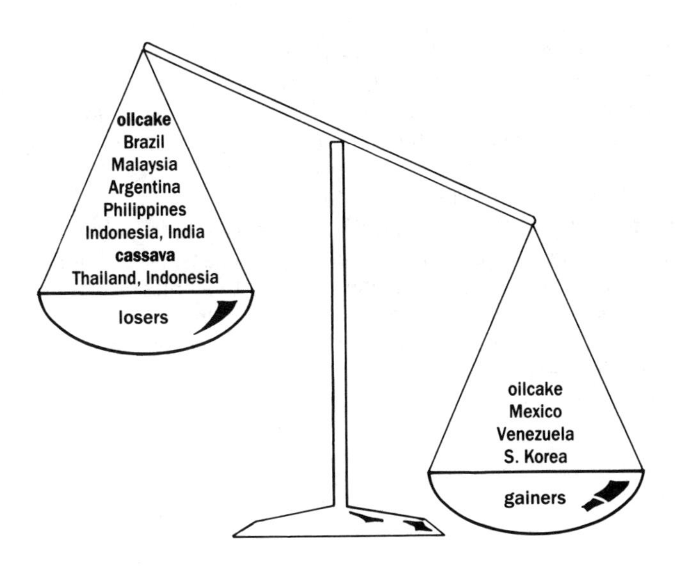

Since many animal feeds are by-products (or co-products) of vegetable oils, any change in either the animal feeds or the vegetable oil market will affect the other. Both producers and importers of vegetable oils would therefore be affected at one remove.

Conclusions

The Panel accepts that the evidence suggests that any of a number of possible changes in the CAP could lead to lower imports of animal feeds. If that happened a lot of people in the Third World would get hurt; they would lose their markets and thus their livelihoods.

The Panel believes that the aim of greater self-sufficiency in vegetable proteins is both costly to the EC and harmful to poor producers in the Third World. It would be better to offer the market to those who need it — and at the same time save money. Farmers in the EC would be given incentives (e.g. cash for conservation) to ensure that they did not lose out by not growing what is better imported.

If changes to the CAP are to be introduced which have the effect of reducing the EC's imports of animal feeds, steps should be taken to identify people in the Third World who would be harmed by this.

The Panel noted the precedent set by the case of the cassava industry in Thailand, where the EC has attempted to make good the damage its policies have caused. It considers that the principle of making good should be applied wherever changes to the CAP can be shown to hurt poor people in the Third World. The STABEX scheme of the Lomé Convention (the EC's trade agreement with 66 mainly small and poor ex-colonial Third World Countries — see Box 1.B, page 13) could cushion the effects of changes on its member countries — but they contain only 10% of the Third World's poor, and ACP countries produce only about 2% of the animal feeds that the EC imports.

One form of making good should take the form of help to develop production of staple foods. Another should be to develop appropriate local Third World animal feed industries, together with help in using the output appropriately to boost local consumption of meat, eggs or dairy products. Box 2.E describes some of the work on this currently under way.

The best way of helping Third World countries to increase their incomes would be to encourage them to do more processing of their agricultural products. This would involve reducing or cutting the tariffs on processed goods. Since this would have an effect on employment in the EC, however, it would be important to analyse the effects very precisely in advance, and to devise ways of minimising them.

Box 2.A

EC IMPORTS OF ANIMAL FEEDS — 1984

	Total imports*		% of total tonnage from Third World	from Third World	
	'000 tn	*£m*		*'000 tn*	*£m*
Soyabeans	16,434	2,708	53	8,697	1,424
Cassava	5,257	490	100	5,254	490
Molasses§	2,236	124	84	1,867	103
Wheat offals etc	1,196	103	86	1,033	90
Citrus pulp	1,417	149	61	860	83
Copra	633	80	100	633	80
Sunflowerseed	952	104	65	604	70
Palm kernel	651	70	100	651	70
Fishmeal	608	182	71	429	62
Linseed	638	92	66	420	61
Cotton-seed	432	60	96	413	57
Rapeseed	582	66	51	295	33
Maize germ	1,036	119	26	271	29
Groundnuts	204	31	86	176	27
Maize gluten	3,734	412	3	125	14
Other	1,783	213	46	817	80
Total	**37,793**	**5,003**	**60**	**22,545**	**2,773**

* imports of oilseeds are taken to be meal/cake equivalent at × 0.8 for soyabeans and various equivalents for all other oilseeds.

§ 75% of molasses is taken to be used in animal feed

Source: Jan Klugkist, *De Veervoerexport van Ontwikkelingslanden nar de EG* NIO Association, Amsterdam, 1986.

Box 2.B

WHAT ANIMAL FEEDS ARE

The original animal feed was grass, growing or dried. Still today, about 58% of the EC's animal feed is supplied by grass (grazing and hay). But the current level of animal production would not be possible without supplementary feeding — raw cereals (about 23% in the EC), and compound feeds.

Box 2.B cont.

The contents of these compound feeds are computer controlled to produce the required mix (which varies for different animals and purposes) using the lowest cost ingredients. Because the price of EC grain is higher than that of imported alternatives (see main text), EC animal feeds contain a high proportion of imported products, particularly those used for ruminants.

Some notes on the main ones;

Soyabeans and meal provide nearly half the EC's imports of animal feeds. The biggest producer is the USA with 60% of world production (57 million tonnes in 1985/6), followed (a long way behind) by Brazil, Argentina and China (where a lot is used directly by humans). Like other oil-seeds, soyabeans produce both oil and cake/meal. Currently the cake/meal produces about 60% of the bean's value. Soyabeans are used in animal feeds for all types of animal.

Cassava is a tuber rather like a large dahlia, which is used as human food in many parts of the Third World. Though a low-grade human food, it can grow on poor soil with little water. Its use as animal feed took off in the early '70s when two EC companies realised its potential as a cereal substitute — given the high price of EC grain and the low level of protection on imported cassava. It has a high starch content and is used in all animal feeds. The main supplier is Thailand — see Box 2.D; Indonesia also exports some.

Molasses is a by-product of sugar. It can be produced either from beet or cane. Molasses is mixed with all types of animal feed, especially for ruminants; it is a good source of carbohydrates and it makes the feed palatable. Main Third World suppliers to the EC; Pakistan, Thailand, Sudan, Mexico.

Wheat offals consist of the outside of the wheat grain when it is milled to produce white flour; milled wheat produces 20–25% offals. It is a useful source of protein (14–16% — compared to whole cereals at 11%). The levy on offals has been increased to protect EC producers and imports are declining. Main Third World suppliers to the EC: Argentina, Indonesia, Nigeria, North Africa.

Citrus pulp consists of the skin and pips of citrus fruits; it is a by-product of the industry which produces fruit juices and canned fruit for human use. It is used in cattle feed. Main Third World suppliers to the EC: Brazil and Morocco.

Copra-cake is made from the flesh of the coconut; the meal/cake is the residue after the oil has been pressed out. Coconut palms are usually owned by small-holders, but the processing is done mainly in large plants. Copra-cake is largely used in animal feed for ruminants.

Box 2.B cont.

Sunflower seeds produce a meal that is rather high in fibre, but which can be used for all livestock. EC production is increasing, especially in France and Spain.

Rapeseed is the main oilseed product of the EC, because it is the one which grows best in a temperate climate. Its yellow flowers are becoming a familiar sight in May and June. It is used almost entirely in cattle feed.

Fishmeal is made mostly from fish not suitable for human consumption. It is high in protein and very beneficial in pig and poultry foods, though usually expensive compared to soya-meal. Main Third World suppliers to the EC; Chile, Peru.

Cotton-seed is a by-product of both cotton and cotton-seed oil production. The oil is often used locally and the residue (in the form of cake or meal) is exported. It is used mainly for cattle. Its use in the EC is declining.

Groundnuts are also a declining component of the EC's animal feeds; they are readily contaminated by a carcinogenic excretion from a fungus, called aflatoxin, and their import is restricted by EC health regulations (although de-toxification plants are available).

Maize germ is the residue that is left after the maize has been crushed to make maize oil. It is used mainly in ruminants' feed. Main Third World suppliers to the EC: South Africa, Argentina, Brazil.

Maize gluten feed is a by-product of the USA's starch and isoglucose industries. Because of the EC's high cereal prices, it fetches a better return in the EC than domestically. It is used mainly in feed for ruminants. Hardly any comes from the Third World.

Source: Brian Rutherford, *Proteins and other feedingstuffs*, paper at Trade Education Course, 1986

Box 2.C

THE ANIMALS THAT EAT ANIMAL FEEDS

EC usage of animal feeds — 1983/4

	Million tonnes	*£ million**	*% of value*
Cattle & calves	29.6	3,700	36
Pigs	26.5	3,312	33
Poultry	21.3	2,662	26
Sheep, horses, rabbits, etc	3.7	462	5
Total	**81.1**	**10,137**	

* based on average price of £125/tonne

Source: FEEFAC, quoted in *Digest of Facts & Figures 1985/6*

Box 2.D

CASSAVA IN THAILAND

The EC's cassava trade with Thailand is an example of where the EC has (albeit inadvertently) created a market and given it to those who need it (in this case, three million very poor people). Because of the situation in Thailand (lack of land rights, declining fertility, erosion, etc), cassava production is fragile enough. An Agreement with the EC at least recognises the EC's responsibility and enables a sizeable level of export to continue.

Cassava is one of the animal feed components that is imported into the EC on a large scale because of the CAP. Four parts of cassava plus one of soyabean meal produce a mixture with the nutritional value of maize. There is a 6% import tax on cassava, and 0% on soya-beans. So the price of cassava-soya mix is lower than either imported maize (which has higher import tax) or domestically produced cereals.

Thailand was well placed to take advantage of this situation. In the early '60s population growth and political immigrants in North-East Thailand meant a lot of people were looking for something profitable to grow. An access network of political/military roads had been built, largely with US help. Cassava grows well on the poor soils. Since it is not a local food (rice is) it can all be exported. And, crucially, its peak labour requirements fit in well with rice cultivation.

The result was a rapid growth in Thai exports — from ¼-million tonnes in 1960 to nearly 7 million in 1984. Cassava now ranks fourth in Thailand's exports. The EC also imports small quantities from Indonesia, China, India and Brazil, but Thailand holds 95% of the market.

In Thailand about 1.5 million hectares are under cassava, some 7% of the total agricultural land area. About 20 million tonnes of cassava are

Box 2.D cont.

produced annually (it takes 2½ tonnes of cassava tubers to make 1 tonne of the exported pellets). Most (95%) is grown on small farms (less than 6½ ha).

Including those involved in transport and processing, about 3 million people depend on the cassava trade. They do well out of it; they get 50% of the Rotterdam port price. And it makes up a lot of their income — 40% for the average farmer, up to 80% for the poorest who have fewer other sources of income.

But the Thais do not cultivate cassava on a long-term sustainable basis. Permanent cropping gradually takes out the nutrients from the already bad soil, and it takes up to 20 years of fallow to recover. As a result yields fall steadily after three years of crops — from 15–25 tonnes/ha, down to a minimum of 6 tonnes/ha. Since few farmers have legal rights to their land, they have no incentive to improve it — to reduce erosion or restore fertility.

Although the cassava trade suits EC animal farmers (who get cheap animal feed) and the Thai farmers (who get a good income), it does not suit the Commission (which has to pay for buying and storing surplus grain which could be fed to animals). In the late '70s pressure to limit the imports grew. The EC could not tax cassava imports because of the GATT (see Box 6.A). Instead, in 1982, it agreed a 'Voluntary Restraint Agreement' with Thailand (and other suppliers also). Under this, Thailand may export 4.5 million tonnes in 1985 and 1986, plus 450,000 tonnes spread between the two years.

Thailand was not at the time a member of the GATT (see Box 6.A, page 70) though it has since joined. So the EC was not compelled to give compensation. However the Agreement also included an aid package designed to promote diversification and to spot new markets both for cassava and for any new crops. The Agreement is renewable from the end of 1986 in 3-year periods at the 1985/6 quantities. It is estimated to cost Thailand about £100 million/year in lost exports. The aid package is about £32 million over five years — though there is some discussion about how much of this is truly 'additional' aid — i.e. aid which would not have been given anyway.

In the face of this Agreement, the Thai Government is caught between encouraging farmers to diversify, and maintaining their income from cassava. For the moment production continues to rise.

Source: *Tapioca from Thailand for the Dutch Livestock Industry,* Andre A Van Amstel, Els E. M. Baars, Jos Sijm and Huub M. Venne, translated by Paul Bruijin, IDES Research Report No. 20, Free University of Amsterdam, 1986

See also: Study of Alternative Markets for Thai Tapioca Pellets and Sorghum, Dr. R. Wollfram and Dipt Ing agr G. Jeub, MBR Consultants, Brunswick, W. Germany, for the European Commission, 1984.

Box 2.E

ANIMAL FEEDS IN THE SOUTH PACIFIC

Many islands in the South Pacific are trying to develop their very small-scale pig and poultry industries. To do this, they import compound animal feeds from Australia and New Zealand. There are three disadvantages to this policy; the imported animal feeds are expensive — they deteriorate in transit — and they are made from raw materials in the islands that consume them.

The British Government's Tropical Development and Research Institute (TDRI) has made an analysis of what is needed to get out of this situation — to enable islands such as Fiji, W. Samoa and Tonga to convert their own agricultural products into animal feeds — and feed them successfully to the pigs and poultry.

Copra-cake, cassava and palm-kernel cake are used on a small scale at present. Their use could be increased. The juice of sugar-cane, grown extensively in Fiji, can be used for animal feed — as work by the TDRI in the Caribbean and Latin America has shown. The false peanut (*cassia tora*) grows like a weed on some of the islands (especially Vanuatu); the TDRI has also worked out how to process, detoxify and feed it to poultry.

The TDRI has developed small-scale machinery which can process these raw materials into compound feeds — and ways of amplifying machinery that already exists in the islands. What is needed in addition is training, quality control, and technical advice. Currently the main source of advice is commercial companies in Australia and New Zealand; most countries feel 'technically exposed' when dealing with them — and lack of local expertise leads to many unsound animal feeding practices.

The islands of the South Pacific are smaller than most countries, but the same techniques for making local use of raw materials to produce animal feeds could be used in many other countries.

Source: TDRI

Case-box 2

THE SOYA WORKERS — BRAZIL

Brazil is the second largest world producer of soya beans after the USA. It is the largest exporter of soya meal and the second of soya-oil. Production in 1984/85 was 17.4 million tonnes from 9.5 million hectares, with the export of 789,300 tonnes of oil, 8.1 million tonnes of meal and 3.5 million tonnes of beans.

In the late 1960's the Brazilian government encouraged small

Case-box 2 cont.

farmers to plant soya under the slogan *'Plant, and the Government will guarantee your produce'*.

Each year they planted a larger area until many of them were producing almost 100% soya. Then came a combination of falling productivity through continuous monoculture, lack of managerial ability, falling product prices and increasing prices for herbicides and fertilisers, plus, more recently, drought. As a result, many small farmers are keen to find an alternative crop.

* * *

In southern Brazil Dona Teresa (52) and Sr Silva (55) have been living on their present plot of 12 hectares for the last 20 years. They have five sons between 12 and 30 and a daughter of 26. The sons live at home — the daughter is a teacher and lives in the town about 5 km away.

"In the old days we had a mixed farm — maize, rice, potatoes, cassava, pigs. Things were easier then. And (the son adds) there wasn't so much soya about". Three years ago they got a loan to buy lime to try to restore the land — but they have not been able to pay the debts. One of the sons added, *"Our hope now is the wheat. We've planted (also with a bank loan) 7 of our 12 hectares with it"*.

The day's work begins between 6.30 and 7 am, breaking for lunch around 11 am and continuing till evening. "*Small-holders don't have holidays*", they say. They mainly cultivate with oxen, from time to time borrowing a neighbour's tractor. With their sons at home they don't need to call in extra help. The two younger boys go to school in the town in the afternoon (a half hour walk away). It costs only £1 a year each. Books are extra.

Water from their own spring is pumped into the house. They have had electricity for the last five years and they have a fridge, TV and radio. The bill varies between £1 and £8 a month. Like most of their neighbours, they have a gas and a wood stove.

Clothes, as one son comments, are expensive: *"You need CZS 500* (£25) *for a bad shirt and jeans"*.

With no free health services, the threat of medical expenses hangs over them. Fortunately a revolutionary municipal health programme is being started nearby. It aims to provide free medical care, and training for health agents. If it succeeds it will greatly relieve their worry.

In this part of Brazil there are not the usual health problems due to poor nutrition. But, as in many other areas there are health problems resulting from chemical sprays.

The elder son, who has tried his fortune in the new frontier areas of Mato Grosso and Goias comments *"There's never a lack of work to do here — we can complain, but you just go and try to earn a salary outside — deep down this life is better"*.

Source: Frances Rubin, Oxfam Recife

Photo: Frances Rubin

Senhor Silva and his family.

3 WHEAT

The Evidence

Most of the wheat traded internationally is used for food, while most of the coarse grains (maize, sorghum, barley, etc) are used for animal-feed. Since it is wheat that forms by far the greatest part of international trade — and which causes most of the difficulty and controversy, the Panel concentrated on wheat.

Viewed from inside, the EC's exports of wheat seem very important. But in a world context, despite sharp rises in recent years it accounts for less than a fifth of world exports. Table 3.1 shows that the USA with 30% and Canada with 20% both export more.

Table 3.1

World wheat trade, 1985/6

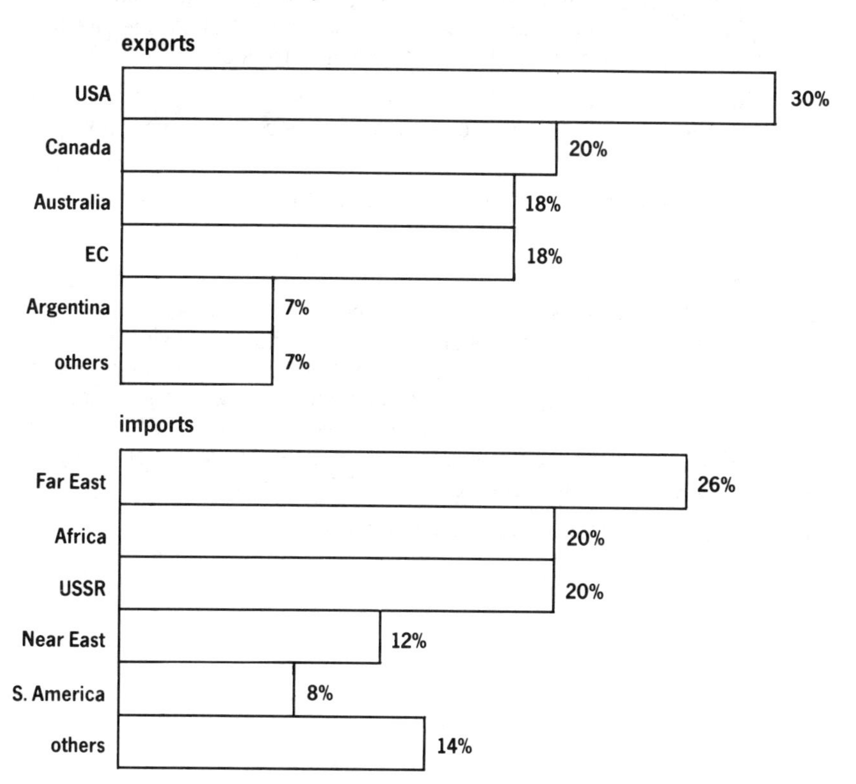

Source: Evidence to the Enquiry, John Ellis, page 83

Table 3.1 also shows that the biggest single importer is the USSR, with a fifth of world wheat imports in 1985/6. Developing countries form by far the biggest category; they take over half of all wheat imports. Box 3.A shows the main Third World importers.

The EC's surpluses do not look so excessive on a world scale. The EC produced and imported some 14 million tonnes of wheat more than it consumed in 1986 — about 22% of its own total production. But the surplus amounted to only about 3.6% of world production.

The World Bank estimates that Third World imports of food (mainly wheat) will continue to increase — perhaps to double their present level within 10 years. Many Third World incomes are rising; as they do so, the tendency is for both governments and individuals to spend more on food.

But even when supplies are short, people tend not to eat much less. This means that variations in supply cause quite marked variations in price.

The four main influences on the world wheat market are:

— **The USA's 'Loan Rate'**, which is similar in concept to the EC's 'intervention price' (see Box 3.B). It tends to set a floor to world prices when wheat is in surplus — as it is now. The 1985 Food Security Act will sharply reduce the Loan Rate and world prices are likely to fall in consequence.

— **The USSR's purchases.** The USSR's import needs vary a lot — by more than all the imports of Third World countries combined. It does not provide a constant effect on prices.

— **Currency instability** — fluctuations in the US dollar affect the export prices of other suppliers, especially Argentina.

— **The EC's surpluses.**

As supplier of around 20% of total exports (and rising), the EC is an important influence on the world market — but not the only and not the dominant one.

There is strong pressure within the EC to reduce the amount of wheat produced. Consumption in the EC is static, but production continues to rise. The increase is almost entirely due to increased yields; since 1976/7 the total acreage down to cereals has remained almost static at 27–28 million hectares — but yields for wheat have risen from 3.4 to 5.7 tonnes/ha.

Table 3.2 gives the production and export figures:

Table 3.2

EC WHEAT SURPLUS, selected years 1975/6–1985/6

Million tonnes

	1975/6	1984/5	1985/6*
Production	37.4	76.4	65.8
Imports	7.1	2.7	1.5
Total	**44.9**	**79.1**	**67.3**
domestic usage	37.5	53.2	53
Surplus (production/ imports less consumption)	**7.4**	**25.9**	**14.3**
Exports	9.5	17.2	15.5
Carry-over stocks	8.3	17	19.6

* EC 12

Source: Evidence to the Enquiry, John Ellis

It costs a lot to store the surplus, or to sell it on the world market at the world price (under the CAP, the EC must pay the difference between the EC price and the world price — currently about £65/tonne). In 1985 the EC spent £564 million on storing cereals, and £323 million on subsidising the export of wheat.[15] A further cost, currently worrying the Euro-Parliament's Budget Control Committee, is the falling value of the EC's stocks. The EC values its stocks at the price at which it bought them; since the price of wheat is falling, so is the value of the stocks — but EC figures do not show this.

Forecasts suggest that unless something is done to reduce production and dispose of the surpluses, the EC's cumulative stock of all grains (not only wheat) will build up to 80 million tonnes in five years' time (1991/2) — enough to feed 444 million people on refugee rations for a year. The cost of storing it would be £1,800 million a year.[16]

What may happen

As Box 3.B shows, the US Loan Rate (which effects world prices) is set to come down still further and US export subsidies are likely to contribute to the trend. Since most of the proposed changes to the CAP would result in higher wheat prices on the world market, the next few years, with US policies making for lower prices, are a good time to make such changes.

In the EC the Commission has already begun to try to reduce the surplus. It will probably continue to do so. It may be successful. If it is, the Third World would be affected in four different ways:

Domingo and Christita Sendina are two of the people who depend on the EC as a market. The copra they grow in Baranguay Tinguin in the Philippines provides 90% of their income. It is made into coconut oil and animal feed.

The nine To sisters grow cassava in Thailand. Their only market is the EC, where the cassava is mixed with soybean-cake to make animal feeds.

1 The current surplus

There are many obstacles to putting the current surplus onto the world market. To do so would raise political problems with other wheat exporters — most obviously the USA. The likelihood of a 'grain price war' with the USA would be increased. World prices would fall further, with a resulting extra cost to the EC's budget; the EC must pay the difference between the world price and the EC price, so it would pay more unless its internal price fell by a similar amount. But the Commission seems prepared to face the short-term disadvantages in order to obtain the long-run gains of not having such a large surplus.[17]

In the Third World grain exporters would lose and importing countries would benefit — at least their governments would, and their urban people; governments tend to use imports to give cheap food to their politically active townspeople. The effect of cheap food imports on local agriculture may not be so beneficial — see below.

Table 3.3 shows which countries would gain and lose from higher world wheat prices.

Main short-term gainers and losers if world wheat prices rose

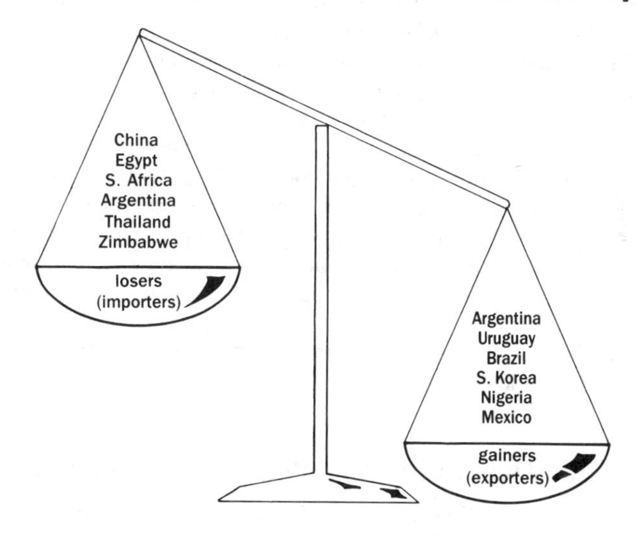

2 A smaller continuing surplus

If there were a much reduced surplus in the future (or even none at all), the immediate effect would be to raise world prices. The effects would be the reverse of putting the current surplus on the world market and lowering prices. In the short term, this would benefit a few Third World countries which export wheat, and hurt quite a lot which import it. In the long-term higher world prices might encourage Third World countries to give their own agriculture a higher priority, and their farmers better prices for their produce. Higher prices do often result in higher production.[18] But while low world wheat prices enable and perhaps even encourage Third World governments to neglect their agriculture, it seems unlikely that slightly higher ones alone would be enough to persuade them to change them.

In any case, in the longer run other producers would probably step in to fill the market and bring the price down again. Some of these other producers might be Third World countries — but the USA would probably take the lead role.

3 Instability

If world production of wheat rises, someone, somewhere must pay less and/or eat more. But even when world prices are low because of rising production, consumers in the EC do not pay less — and are not encouraged to use more or to grow less. So world prices have to fall further before the market clears.

Annual world cereals production has varied much more in the last two decades than in the period immediately before that. The reasons are unclear — but this increased variability has coincided with increased EC cereals exports. Some research suggests that the CAP is responsible for over half this increased variability in wheat prices[19] — though estimates vary, and other factors (such as USSR purchases and protection by other countries) play a part. A continued high variation in production and therefore in prices is likely.

This price instability works very much to the detriment of Third World countries because they are heavily dependent on agricultural markets. Assured grain imports and markets for exports are very important to Third World countries; they are already trying to cope with very variable production (weather, disease, etc) themselves. The combination of local and world variability make effective planning almost impossible. This applies to both imports and exports (it adds to the risks of exporters, and undermines investment).

The CAP's effect on price instability is therefore an adverse influence on the food security of Third World countries.

4 The method by which the EC reduces its production

The Third World may also be affected differently by different methods used by the EC to reduce its surplus of wheat. Box 3.C sets out the main possibilities currently being considered.

If prices were reduced in an effort to cut production, for example, EC farmers might respond in the short term by growing more wheat to keep up their incomes. In that case, no extra land would be available for growing crops currently imported from the Third World. But if homegrown cereals (other than wheat) became cheap enough to replace some animal feed imports, then Third World countries would lose their current markets. If quotas were used to control production, then farmers would probably switch to growing other things, including products currently imported from Third World countries — such as oilseeds. Either way Third World countries would lose.

A suggestion currently being discussed in Britain is to tax or ration agricultural inputs, especially nitrate fertilisers. The argument is that this would reduce production (and perhaps costs), and also pollution. Opponents of the idea emphasise the administrative problems and the scope for abuse. In addition they point out, new varieties giving yields with less nitrogen would probably soon be developed. From the point of view of the Third World, the proposal is attractive because it does not involve using land taken out of cereal production for another crop which competes with a Third World export.

The Commission is currently investigating new uses for cereals, such as ethanol (as engine fuel) and starch. At present there is no sign of a commercially viable development — but if there were, Third World countries could lose markets if the new products replaced current imports from the Third World.

Conclusions

The Panel noted that there is disagreement about the extent of the EC's influence on world wheat prices; changes in the CAP would clearly affect prices in the short-term — but perhaps not much in the long-term. The Panel also noted the experts' disagreements about the effect on the Third World poor of higher wheat prices; while farmers and long-term agriculture could perhaps benefit in the longer term, some of the very poorest people of all, in the towns, would suffer for a perhaps lengthy short-term. So the Panel concentrated on issues where helpful action can be seen more clearly.

Whether or not Third World countries benefit, the pressures to reduce EC wheat production are sure to continue. The Panel considers it important that each proposal to reduce wheat production should be checked for its effect on the Third World; this particularly applies to alternative crops, or alternative uses for existing crops.

The exact extent of the EC's effect on world price instability may be in doubt — but its unfavourable effect on Third World countries is clear. Some way should be found to reduce this instability — probably by linking EC production and consumption more closely to world prices.

The Panel received two suggestions to use food aid programmes to solve some of the problems created by the over-production of wheat — one to compensate for price instability and one to get rid of current EC surpluses. Though attractive on paper, the Panel felt that these proposals would be unlikely to be productive, given the general limitations on the productive use of food aid (see next section).

Box 3.A

MAIN THIRD WORLD WHEAT IMPORTERS, 1983/4

Third World countries importing more than 1 million tonnes wheat equivalent (including flour)

Million tonnes

China	9.8	Algeria	2.4
Egypt	7.3	Morocco	2.3
Brazil	4.3	Nigeria	1.7
Iran	3.6	Cuba	1.7
Iraq	3.0	Bangladesh	1.6
India	2.5	Indonesia	1.6
S. Korea	2.5	Chile	1.0

Source: International Wheat Council

Box 3.B

THE US SYSTEM OF AGRICULTURAL PROTECTION

In the USA about 80% of farmers take part in a voluntary price support scheme for such products as wheat, cotton and rice. Here is how it works for wheat:

— the farmers agree to take a percentage of their 'base agricultural land' out of production (they can choose the least productive). The percentage is currently 27.5% for wheat.

— they are entitled to put their harvest under a 9-month 'loan' to the federal Commodity Credit Corporation (CCC). After 9 months, farmers take their harvest out and sell it. They can either sell it on the open market — or to the CCC at what is called the **'Loan Rate'**. Sometimes the Loan Rate is higher than the market price (as it is for maize currently), sometimes not (as for wheat currently). The CCC thus provides a guaranteed market for farmers' output and the Loan Rate thus determines farmers' minimum prices — and also the minimum price on the domestic market.

— they are entitled to **Deficiency Payments** to make up the difference between the Loan Rate and the **Target Price**. The Target Price is set by the government to reflect reasonable production costs. The Target Price is often substantially higher than the Loan Rate; the current Target Price for wheat, for instance, is $4.38/bushel, while the Loan Rate is $2.28. The Target Price affects farmers' incomes and government costs, but not the price on the open market.

Box 3.B cont.

Part of the Defiency Payments are made in Generic Certificates. Farmers can either exchange these for government owned food stocks, or sell the actual certificates. This device tends to off-load the government-held surplus onto the market and thus to push prices down.

In addition, the Government pays an **export subsidy** to shippers to encourage them to sell selected products (for example wheat) to selected markets (for example North Africa and the Middle East). This subsidy, currently around $20/tonne, is paid in Generic Certificates. A sum of £1,300 million/yr has been allocated to this subsidy.

The Government can also pay farmers to keep additional acreages out of production — beyond the 27.5% they have already agreed.

The future

In an attempt to cut expenditure and production in 1985 the Government tried to cut both Target Price and Loan Rate over a 5-year period. But Congress froze the Target Price (on which farmers' incomes depend) for three years (i.e. until 1988), allowing it to be lowered thereafter at 5% a year for 2 years. The Loan Rate however (which affects US and world market prices) has been cut — by nearly 20% for maize, and nearly 30% for wheat. The Government has authority to cut it further; there are indications that it will be cut by an additional 20% over the next two years.

The effect on world prices

By linking Deficiency Payments to reductions in acreage, the US system to some extent prevents higher prices from encouraging production — which would lower domestic prices. On the other hand, the use of Generic Certificates re-cycles some of the surplus onto the open market — thus lowering prices.

The USA is the dominant world grain exporter, and has few ways of insulating its domestic prices from world prices. The minimum US domestic price is set by the Loan Rate. So the Loan Rate also sets the basic price in the world markets. Thus US plans to cut the Loan Rate are likely to result in further falls in world prices.

Sources: Agriculture Department, US Embassy, London. *World Development Report, 1986*, IBRD, Washington, p. 119

Box 3.C

WAYS TO CHANGE THE CAP

A number of ways of changing the CAP are under discussion:
— some want to cut the cost by limiting imports
— others seek to cut the cost by limiting production
— still others want to reduce protection and let in more imports.

These different views are reflected in the various proposals put forward for reform of the CAP:

— **Lower prices to farmers for their products** — would reduce the cost of the CAP and the surpluses. But this method would hurt the very group the CAP was set up to protect — farmers, especially smaller ones.

— **Quotas** — to limit production. Many in the Commission are strongly against quotas because they are very costly to administer and, in the Commission's view, they lead towards 're-nationalisation' and the splitting up of the EC. They are also likely to ossify production patterns and limit opportunities for growth to small-scale farmers. In many parts of the EC this would be particularly serious because of the lack of alternative opportunities.

— **Co-responsibility levies** — to make producers pay for disposing of surpluses. There is a lot of support for this approach in the Commission — but it tends to result in higher prices for consumers, and it assumes that a market for the surpluses (eg. in sugar) exists and is desirable. See Box 5.A (page 59) for details of the sugar co-responsibility levy.

— **Expansion of protection** — leading to the exclusion of dairy products (especially those from New Zealand), sugar (whose cost would be transferred to the aid budget) and beef — and to taxes on animal feeds and vegetable oils. But any moves of this kind would lead to serious confrontations with suppliers.

— **Taking land out of production** — either leaving it fallow for environmental or leisure purposes, or turning it over to forestry. One suggestion is that up to 1.3 million acres could be taken out of production in Britain by the year 2000 — an area twice the size of East Anglia*. But the cost of administering this policy would be high and it could increase the rate at which agricultural jobs are being lost.

— **Alternative crops** — the Commission is desperate not to increase the rate at which agricultural employment is already falling, so it has investigated several crops as alternatives to those producing the current surpluses. It lists major disadvantages for all of them, however — they are either too small-scale (e.g. nuts, medicinal plants, bee-keeping and fish-farming), too long-term (e.g. wood,

Box 3.C cont.

tree-fruits), too expensive (e.g. vegetable protein like lupins), uneconomic for the foreseeable future (e.g. new industrial uses for existing crops — such as ethanol for fuel from sugar beet, cereals, potatoes, chicory or artichokes), or require too much infrastructure (e.g. cotton).

Source: European Community Commission, *Perspectives for the Common Agricultural Policy — the Green Paper of the Commission,* Brussels, July/85

* Simon Gourlay, President of the National Farmers' Union, *The role of agriculture in the countryside — a farmer's view*, paper to the Oxford Farmers' Conference, Jan/86

4 FOOD AID

The Evidence

The modern system of international food aid began, and to a some extent continues, as an aspect of agricultural policy. Right from its start in 1954, US Public Law 480 (PL480) treated food aid as a way of managing exportable surpluses; to this day PL480 is part of the US Department of Agriculture (USDA). Internationally the Food Aid Convention (the agreement which controls food aid) was negotiated as part of the International Grains Agreement (an agricultural trading arrangement). In the EC too, the Development Directorate (DG VIII) shares control of food aid with the Agricultural Directorate (DGVI).

But although agricultural policy continues to play a part in food aid, the programmes now (and perhaps increasingly), are organised in humanitarian and development terms. The EC's food aid policy has these objectives:

— to raise the standard of nutrition
— to contribute towards balanced economic and social growth
— to help in emergencies.

The availability of surplus food has influenced both the scale and the ingredients of the EC's food aid programme; it is seen as a convenient way of reducing the embarrassing surpluses. Nobody starting from the needs of Third World development would have invented food aid on its present scale.

The scale is considerable. Box 4.A shows the world, US and EC totals of the various types of food. Box 4.B shows the main countries which get it. Box 4.C describes some of the international food aid institutions.

The EC's food aid programme did not begin until the late '60s, and it first became a significant resource in the Sahel famine of 1972–4. In recent years it has accounted for between a quarter and a third of all food aid. The USA provides over half; the rest mostly comes from Canada, Australia, Japan and Scandinavia.

Food aid makes the headlines when it is used for emergencies — the sacks thrown out of the plane or the convoys crossing the desert. In fact **emergency** food aid is generally less than a fifth of the total (the recent past is untypical because of the NE African famine). Box 4.D describes the various types of food aid. Case-box 4 describes four rather different real-life recipients of food aid in Bangladesh.

The 'need'

The 'need' for food aid is generally calculated by assessing the nutritional requirements of a country's population, and subtracting an estimate of its supply (production and stocks) of staple foods (maize, sorghum, rice, potatoes, cassava etc). This produces the 'import requirement'. Any imports that cannot be financed commercially are then deemed to require food aid. How much each country can 'afford' to spend on commercial food imports is, of course, open to discussion; countries which decide to spend a high proportion of their foreign exchange on food, have less for fuel, agricultural inputs, spare parts, tools, arms or luxuries. So from the point of view of food aid donors, food aid releases foreign exchange which can then be spent on other products of the rich countries. It is basically a means of budgetary support; the Third World government can spend the money it saves how it wishes — on the tools for development, on cheap food for townspeople, or on arms. The purchases may or may not benefit poor people.

Most forecasts agree that a number of countries, particularly in Africa, are likely to need to import food for at least the rest of the century. This is partly because of their lack of resources (either domestic or aid) with which to raise agricultural production, partly because of their rapidly rising populations, partly because of inadequate agricultural policies, and partly because, as incomes rise, so does the demand for food. If serious and repeated food shortages are to be avoided, therefore, if development is to be possible, these countries will need some help.

Food aid however, may not be the most appropriate form of help. For, aside from its use in emergencies, a number of important criticisms are laid against food aid:

— the key to most Third World countries' progress lies in the development of their own agriculture. Food aid may release foreign exchange for use in improving agriculture; but it is a two-edged tool; unless used very carefully, it may reduce the incentive to farmers to grow more, and to governments to give a high priority to agriculture.

— food aid can lead to the establishment of a taste for imported food which cannot be grown locally (most obviously wheat). This means that there is a continuing pressure to use scarce foreign exchange on buying it commercially. This tendency is well established; US Department of Agriculture states quite openly that it sees food aid as a 'loss-leader' to develop future markets.[20]

— food aid sometimes results in shoddy work when used on development ('Food For Work') projects — and the projects themselves all too often benefit the rich once they are completed.

— the temperate agricultural products of the food aid donors are often inappropriate to local requirements — skimmed milk, and butter for example.

— studies show that when used to give school-children a midday meal, food aid does not help the poorest; the poor do not go to school, or soon drop out; those who stay get less food from their families to compensate.

— food aid is not free. If the basic cost of the actual grain is £100/tonne, the full cost of delivering it by air in an emergency can be nearly £500. Even in ideal circumstances (e.g. bulk grain by ship to a coastal African destination) the delivery cost would add 50% to the cost price. The costs are especially high for dairy food aid — which accounts for a third of the cost of all EC food aid (see Box 4.E).

— dairy food aid poses particular difficulties; it is often used to subsidise a Third World dairy industry whose customers are the middle and upper income townspeople.

Fortunately, not all food aid has negative effects; there are cases of productive use (even of dairy food aid). Fortunately too, there is the possibility of reforms to the EC's programme. The Commission has set up a number of working groups; the reports of two, on procedures and on transport, have not yet been made public. A third, on de-linking food aid from agricultural policy, has led to formal proposals for change.

The current programme has been criticised in several ways:

— it is over-influenced by the availability of surpluses
— it is sometimes inappropriate — money would be more use
— it is very slow; food can be delivered in 6 weeks — the EC's average is 24 weeks.

The British Minister for Overseas Development has recently called for food aid to be 'de-linked' from the surpluses, for Third World countries to be offered a choice between food aid and money to develop their agriculture, and for the whole process of delivery to be speeded up.[21] And the Euro-Parliament is pressing for money to be made available as an alternative to food aid.

In responses to these criticisms, the European Council, Commission and Parliament have been actively considering reforms — such as improved management procedures, and using some funds to buy food from other Third World countries. As the Panel finished its report, a number of regulations were in the process of being passed. Their effect would be to meet some of the Panel's criticisms. They are, however, only regulations; they will need to be translated into practice.

Conclusions

The Panel are strongly against the use of food aid to justify EC surpluses. The EC's problems should not be exported to those even less capable of handling them.

Despite the many criticisms laid against food aid, the Panel believe that it can be useful. Its use for development is fraught with difficulties; it should be used only sparingly and with great care. This particularly applies to dairy food aid, which is both expensive and hard to use to benefit the poor. A lot more thought is needed if the existing surpluses are to be used productively by Third World countries.

The Panel welcomes the idea of Food Strategies. The EC has been encouraging some African countries to work out 'food sector strategies' which the EC then backs with aid — whether in cash or food. Food Strategies are not dependent on food aid — though food aid is often used in them. They provide a potential method of using food aid productively. But their track record so far is disappointing. Again, the Panel believes that a lot more work is needed to make the idea successful.

A Food Strategy is one way of encouraging a Third World country to develop its agricultural policy. Many Third World countries neglect their agriculture — and indeed make use of food aid to do so. So it seems sensible to try to reverse this process. There are dangers, however, in rich countries telling poor countries what to do; for one thing the 'correct' policy is seldom clear (the Third World is littered with failed Euro-American ideas) and for another rich countries sometimes use their power to benefit themselves rather than the poor. The Panel consider therefore, that pressure is best exerted by laying down general principals — and any country complying would be eligible for the appropriate food aid.

The organisation of the EC's food aid programme needs to be improved, DG VIII (Development) should take full control of the programme.

The EC's slow response to emergencies quite shocked the Panel; they believe that the procedures should be radically improved as soon as possible. The Panel agreed with the proposal to keep stocks of emergency food available for quick release, preferably in Third World holding centres such as Singapore or Djibouti.

Box 4.A

FOOD AID

Shipments 1983/4

Cereals (grain equivalent)	*Million tonnes*					
	EC	**% of total**	**USA**	**% of total**	**Other**	**Total**
Wheat & flour	1.40	19	4.40	59	1.60	7.4
Maize, sorghum, etc	.34	26	0.79	61	0.17	1.3
Rice	0.14	13	0.49	45	.47	1.1
Total cereals	**1.90**	**19%**	**5.7**	**58%**	**2.24**	**9.8**
Other products	*Thousand tonnes*					
Dried skim milk	171	48	163	46	22	356
Vegetable oil	16	5	271	80	21	338
Butter oil	56	100	neg	—	neg	56
Other dairy products	3	7	33	80	5	41
Total other products	**256**	**31%**	**467**	**60%**	**29**	**791**

***Source:* FAO, unpublished**

Box 4.B

WHERE IT GOES TO

Main recipients of EC Community Action* food aid, 1984/5

Cereals

Countries§	Tons	Organisations	Tons
Bangladesh	140,000	UNHCR	145,000
Ethiopia	135,000	WFP	110,000
Egypt	120,000	NGOs	98,000
Mozambique	50,000	UNWRA	2,000
Niger	41,000		
Sri Lanka	40,000		
Chad	31,000		
Others	248,000		

Total cereals 1,160,000 tons

Skimmed milk powder

Countries§		Organisations	Tons
India	16,000	WFP	35,400
Egypt	4,500	NGOs	26,400
Tunisia	3,000	other UN	5,000
Ethiopia	2,000		
Nicaragua	1,800		
Peru	1,400		
Others	13,100		

Total skimmed milk powder 108,600 tons

Other dairy products

Countries§		Organisations	Tons
India	3,750	WFP	9,435
Ethiopia	2,800	NGOs	3,150
Egypt	1,900	other UN	2,350

Total other dairy products 28,700 tons

Other products

The EC also gave a total of 54,597 tons of other products, which included haricot beans, vegetable oil, sugar, dried fish, raisins, biscuits and sardines to 15 countries and through most of the usual UN and NGO organisations.

* excludes National Action — i.e. food aid given by individual EC member states

§ includes food aid channelled through UN Agencies and NGOs to specific countries

Source: Food Aid Bulletin, No. 2, April 1986, FAO, Rome

Box 4.C

THE WORLD FOOD PROGRAMME AND THE FOOD AID CONVENTION

The World Food Programme was set up in 1961 as the first multilateral food aid agency. It aims to supply and co-ordinate food aid, and deals with both emergency and development work. In 1983/4 it handled a quarter of all food aid shipments. World Food Programme donations may not be sold on the markets of the Third World countries which receive them.

The Food Aid Convention was adopted in 1967 as part of the International Wheat Agreement. It was designed to meet worries that not enough food might be available for food aid (the USA was restricting the acreage planted). The current Convention, signed in 1980, guarantees 7.6 million tonnes of food aid a year from 22 donors.

Food aid is distributed according to the UN's Food and Agriculture Organisation's *Principles of Surplus Disposal* which are designed to minimise the effect of food aid in discouraging local farmers.

Box 4.D

THE TYPES OF FOOD AID

Surprisingly, most food aid is not used as it is shown on television — to bring dramatic relief to the starving. Emergency food aid is only the smallest of the three main categories of use.

Programme food aid (57% of total food aid in 1984) is used as a way of giving money indirectly to Third World countries; the food is given to the governments, who can then either sell it on the market and keep the proceeds to meet general government expenditure, or use it to feed civil servants, the police and/or the military. Either way, the government saves money — often in the form of foreign exchange because it would otherwise have had to import the food.

Project food aid (28% of the total in 1984) is used to promote development. Irrigation, roads, terracing or similar projects are financed by paying the local labourers in food instead of money. Project food aid is also used for school meals or in mother-and-child health clinics.

Emergency food aid (15% of the total in 1984) is used to meet short-term immediate food shortages or famines.

Sources: Against the Grain, Tony Jackson & Deborah Eade, Oxfam, Oxford, 1982
Food Aid in Figures, No. 3, 1985, FAO, Rome

Box 4.E

THE EC's FOOD AID

Community Action* 1986

	£ million
Cereals	191.4
Milk products	148.6
Veg oil, pulses etc	27.4
Other	75.3
Total	**£442.8 million**

* excludes National Action — i.e. food aid given individually by EC member states

Source: DG VIII, European Commission

Case-box 4

FOOD AID IN BANGLADESH

Some food aid goes to the very poor. And some goes to governments to spend more or less as they like (see Box 4.D). The poor may benefit indirectly because the government saves on salaries and/or foreign exchange. But the immediate beneficiaries are not necessarily poor.

This case-box describes four beneficiaries of food aid in Bangladesh.

Mohirun

Mohirun is surely the ideal sort of person to receive food aid. She is a widow of 50 years. Her husband died fourteen years back. Since then, she has been managing the family with 2 sons and 6 daughters. Now she is living with her 16-year-old youngest daughter, Jharna. Mohirun failed to arrange a marriage for her because she is too poor to give a dowry.

Her two sons are married and are living apart from her. They refused to take responsibility for their mother and sister. They are working as wage labourers. Mohirun said, *"My sons told me that they did not inherit any property from their father. From their income they are unable to feed their own wives and children. How can they feed us?"* Mohirun's five other daughters are married.

***"I have no land, not even a homestead. The small amount of land that we did have, had to be sold at the time of my daughters' marriage. I could not keep the land."* Mohirun is now living on the verandah of one of her uncles, a medium sized farmer and in the rice business.**

Working on the Road Maintenance Programme.

Case-box 4 cont.

"I am working as a labourer in the Road Maintenance Programme for the last two years. I am one of a group of 15 women. I work from 8 am to 1 pm. The work is available for 14 days in a month and we get paid at the end of the 14 days. We get paid at the rate of Taka. 12 (27p) per day." An agricultural labourer in Mohirun's area earns at least Tk.20 (45p) per day.

When Mohirun works on the Road Programme, it takes her 2½ hours of labour to pay for her staple food of the day, one hour to pay for pulses and the other 1½ hours to pay for salt, spices, oil etc. She also buys firewood, as well as spending another hour or two collecting it.

During the 16 days a month when there is no Road Programme work, she works from morning till evening (10 hours) in her uncle's house just to get food and shelter.

Mohirun used to receive wheat under the Vulnerable Group Feeding (VGF) Programme run on a card system by the Union Parishad (UP — local government authority). She used to receive 35 kg of wheat in the beginning; gradually it was reduced to 25 kg. After 3 years it was stopped. She has heard that the UP has issued newer cards to others. She doesn't get the VGF Ration any more.

"A member of the UP told us that if wheat comes then we will

Case-box 4 cont.

automatically get it. We are still waiting for that. He is giving wheat to others."

VGF wheat is sometimes given to women on the condition that they accept sterilisation. Distribution of the wheat is in the hands of the Union Parishad. But the UP is also interested in achieving its target of sterilization. If it can successfully fulfill its target, then it receives an award of Tk. 50,000 (nearly £1,200); if it fails, it is considered as a bad performance. So the UP chairman and Members use the VGF wheat to force the poorer women to accept sterilization.

Mohammad Halim — Army Ration recipient

Mohammad Halim has been working in the Army for the last 15 years. He is currently at the rank of Habildar, a non-commissioned officer. He earns a total salary of Tk.2,300 (£52.30) per month. He has a house at Tk.30 (68p) rent per month. He also has an Army Ration card for four persons in his family. An Army Officer like Halim is entitled to get cards for only immediate family members — i.e. his wife and children. There is no limit on the number of children.

Halim receives the following items at a subsidized rate from Rations per month:

	Quantity/kg	Price/kg	Open Market price/kg
		Tk	*Tk*
Rice	30	2.24	12.00
Flour	25	1.98	8.00
Sugar	8	6.61	26.00
Soyabean Oil	5.5	8.30	30.00

The quantity of the items received from the Ration is more than enough for Halim's family requirement. He can sell some of the oil, sugar etc, to earn an average of Tk. 70.00 to Tk. 100.00 (£1.60–£2.30) per month.

"Without the Rations I could not have managed. I am not complaining of my low salary because of the privilege of Rations I get from here", Halim said.

In the Army they are entitled to get Rations even after retirement. But the price is double.

Farhana Ahmed — Public Ration Card recipient

Mr and Mrs Ahmed live in an apartment house. They enjoy the luxury of amenities of modern life such as colour TV, refrigerator, etc. They too receive a share of food aid.

Case-box 4 cont.

"If we did not receive any Ration, we would have to spend a bit more money on food items", said Mrs Farhana Ahmed. She is holding a Ration card for four members of her family. Her husband is a government employee — a Civil Engineer working in the Water Development Board. Both husband and wife are working. The net monthly salary is Tk. 6,200 (£140). They have other sources of income, too. For example they bring all the rice they need from their own land in their village home. They also own two plots of land in Dhaka on which they are going to build a house in the near future.

Mrs Ahmed is entitled to receive wheat, rice, sugar, oil (soyabean) and soap from the Ration. But she takes only wheat, sugar and soap. She gets them more cheaply than she could buy on the open market. But she is unhappy about the quality of the wheat and sugar which she gets from the Ration. She said, *"If we could get the items at the price and quality that the Army officials are getting, we would be quite satisfied even with a lower salary."*

She does not take rice from Ration because the quality is very poor. Friends of Mrs Ahmed take this rice *"only for the servants"*. It is too expensive for them to feed the servants the same rice that the family eats. Some of their friends give the whole Ration card to the servants. The servants are not eligible to get a Ration card on their own under any criteria.

Sazzad Hossain — Police Ration recipient

"I am a policeman, not a civilian," said Mr Sazzad Hossain. *"There is definitely a difference in the privileges given to us by the government"*. He is getting a monthly Ration of the following items:

	Kg
Rice	20.09
Wheat	25.70
Pulses	7.47
Sugar	1.87
Oil (mustard)	1.87
Oil (soyabean)	1.87
Other vegetable oil	1.87
Firewood	74.76

He pays Tk.90 (£2.04) per month for these items as a package. The market price for them would be Tk.900 (£20.45). That is, he pays one-tenth of the market price.

Case-box 4 cont.

Mr Sazzad has a family consisting of his wife and a daughter. In the Police Ration system, they get cards for only one child.

Mr Sazzad earns a monthly income of Tk.1,200 (£27.30). He also has a house, for which he is paying Tk. 30 (68p) per month. He also pays for gas, electricity and water.

"Without the Ration, I would not be able to manage my living standard" Sazzad said. He said that the higher officials receive better quality Ration items. This is their privilege.

Source: Development worker, Bangladesh

5 SUGAR

The Evidence

Sugar provides the most striking example of the effect of the CAP on the Third World poor. Because sugar beet, which can be grown in the EC, and sugar cane, a tropical crop grown in many Third World countries, are interchangeable from the consumer's point of view and span much the same range of production costs, the EC has the technical ability either to grow all of its own sugar needs (or more) — or none. Even a minor move in either direction would affect the lives, and sometimes even the deaths, of the thirty million or so people in the Third World who are dependent on cane sugar cultivation.[22]

But sugar is not a high priority for reform in the CAP;

- there are no huge and unmanageable surpluses as there are in other products
- the arrangements for subsidising sugar are to some extent self-financing and self-adjusting — these arrangements are described in Box 5.A
- there are few management problems (as there are with milk quotas, for example).

Change is therefore less likely than for other products — but if it came, it would have important effects.

The world scene

Box 5.B shows how much sugar is grown, how much is beet and how much cane. Less than a third of total production is traded internationally — and less than a fifth is traded on the 'free' market (i.e. outside Special Arrangements like the Lomé Convention (see Box 5.C) and the USSR-Cuba agreement).

Sugar cane is often thought to be cheaper to produce than beet — if only because of cheap Third World labour. In fact, as Table 5.1 shows, production costs vary widely for both cane and beet, with neither being consistently cheaper.

Table 5.1

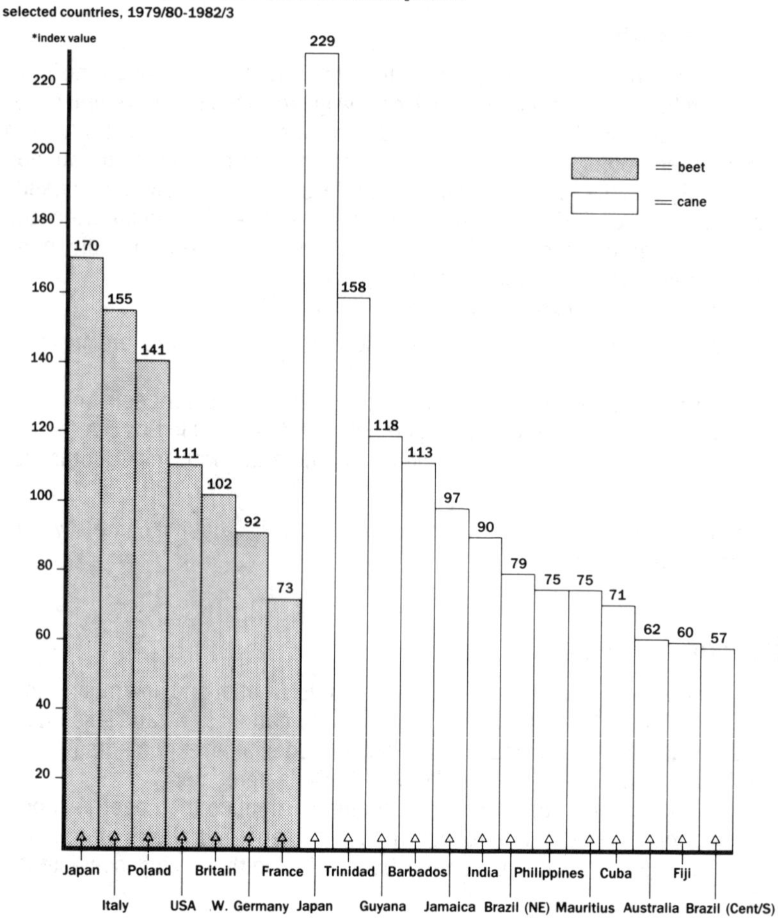

Source: Evidence to the Enquiry, Simon Harris, Table 3, see page 85

Basically because virtually all governments protect their sugar industries, there is now a built-in tendency to produce and process more sugar than consumers can or will buy. World sugar processing capacity is now about 115 million tonnes/yr — while consumption is only about 96 million tonnes/yr. Although consumption may rise a bit in the future, so will production — many Third World countries have plans to expand their sugar industries. Third World sugar producing countries aim to avoid imports by growing and processing all the sugar they need internally, and to get some foreign exchange by exporting the surplus, even at very low prices.

In general the protection policies of the rich countries have very adverse effects on Third World countries. The World Bank reports estimates that Third World countries lost export revenues of some £1,400 million and increased price instability by about 25%.[23]

In the sugar world, the EC is seen as both villain and saint. It is widely blamed for causing the current low sugar price which reduces Third World countries' income. In fact, although the EC is a major exporter, it is not the only culprit, and it is difficult to allocate international blame for low prices. Probably the EC is blamed more because it is a late-comer; in the last 20 years it has grown from a neglible exporter (1% of world exports in 1970) to the major one (22% in 1985). It also obstructed attempts to set up an effective International Sugar Agreement in 1977 and it was closely involved in the disagreements which led to the collapse of negotiations for a fifth.

But the EC has also inadvertently become a world leader in 'altruistic' trade. The Sugar Protocol of the Lomé Convention gives 16 Third World countries the opportunity to sell 1.3 million tonnes of sugar at near CAP prices. When the Protocol was negotiated the world price was three times higher than that in the EC, so the Protocol offered merely a guaranteed market. More recently the EC price has been way above the world price the Protocol member countries would otherwise get; in 1986 the world price was £6–£8/tonne — the price the EC paid was around £26/tonne. The details are shown in Box 5.C.

Since the EC already grows more than enough for its own consumption, it re-exports the equivalent of the whole of the amount it imports under the Protocol, plus as much again of its own surplus — at a huge loss (nearly £2,000 million for the four years 1982–1985).[24] The EC now gets no benefit from this arrangement. The Protocol has become, in effect, an aid programme, recognising the responsibility of several EC countries for setting up Third World sugar industries in colonial times, and transporting to sugar-growing areas more people than the land could otherwise support.

The Protocol however does have problems; it divides Third World sugar producers into two camps — those with access to the EC and those without. This makes it more difficult to negotiate an agreement which would benefit **all** poor sugar-workers. And, because some Third World countries pass on the benefits of the Protocol price by paying their producers above the world price, they encourage them to produce sugar beyond the amount allowed into the EC — which therefore contributes to depressing the price of all Third World countries' sugar exports.

The effect on the Third World

At least thirty million poor people around the Third World are dependent on sugar cane for their survival — probably more than for any other CAP-related product. The countries in which they live are not those which produce or even those which export the most. Some 40% of the world sugar trade is carried out under protective agreements (the EC's Sugar Protocol, the US-Phillipines agreement, the USSR-Cuba agreement, etc). So the people who do least well out of the sugar trade are those who live in countries with a high proportion of unprotected exports. Table 5.2 shows which they are.

Table 5.2

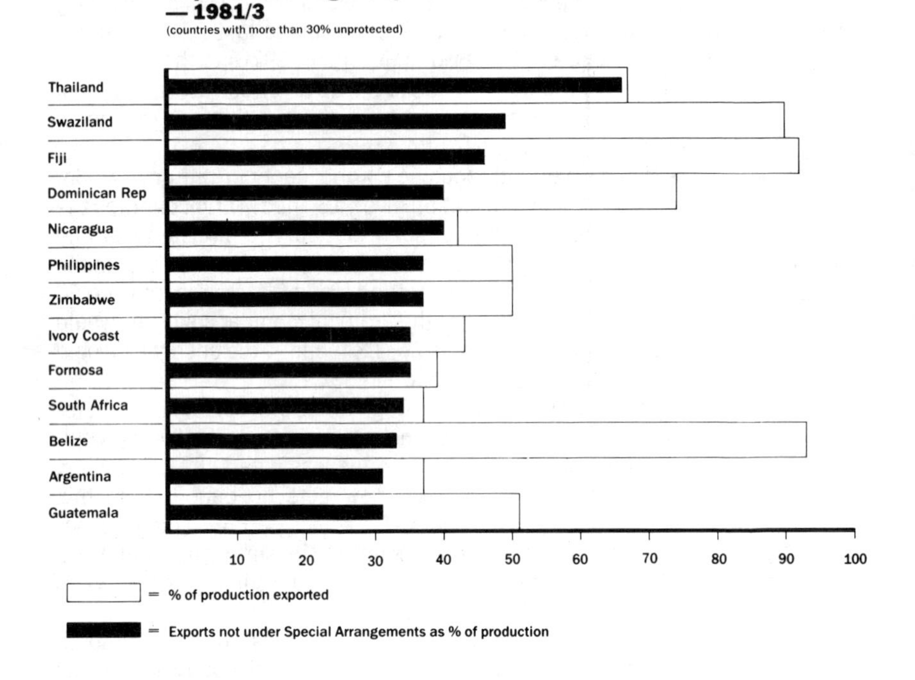

Source: Simon Harris, *After the ISA*, Address to *Outlook '85*, Washington, Dec/84

What may happen

Most of the things that may happen over the next few years would be potentially damaging to poor people in the Third World;

— **Consumption may continue to fall.** Consumption in the EC, the USA and Japan has been falling for over a decade — by nearly 4½ million tonnes since 1973, about 5% of world consumption. It is not clear how far this trend is due to changing diets based on health fears, or CAP-type policies which keep prices high and encourage the use of non-sugar sweeteners. Either way the trend may well continue.

— **The use of alternative sweeteners may continue to grow.** The main alternative sweetener is High Fructose Corn Syrup (HFCS), produced from starchy plants like maize, cassava or sorghum. It is most used in the USA where it has displaced sugar in soft drinks and is widely used in processed foods and bakery products. As a result the USA has reduced its imports from 5 million tonnes in 1981 (half of its consumption) to 3 million in 1983 and possibly to 1½ million in 1986.
Sugar interests in the EC have managed to restrict the use of HFCS to 1.6% of total sweetener consumption. Both EC beet growers and Third World cane suppliers benefit from this.
Some Third World countries (eg Argentina, Egypt and Uruguay) are starting HFCS production plants of their own. Others may follow suit. Meanwhile, non-calorific sweeteners such as Aspartame increasingly appeal to health-conscious consumers in the rich countries — where most sugar is consumed. There is a strong possibility therefore, that while the demand for sweeteners may expand, the demand for sugar may fall.

— **Many Third World countries have plans to expand their sugar industries** — for example China, Indonesia and Cuba. Others plan to regain former production levels — for example Mexico, Chile, Nicaragua. Third World countries boost their industries both to create employment and to save or make foreign exchange.

— **Price 'peaks' may become less frequent and lower** as a result of these factors. Traditionally sugar prices rise and fall in cycles of about 6 years and the price peaks provide the money to re-invest and develop. With less of this money, the weaker industries will find it hard to modernise, and production will be further concentrated in the stronger industries.

— **The trend towards white sugar may continue**. If it does, it will increase the vulnerability of exporting countries with small-scale industries who cannot justify the installation of white sugar processing facilities.

— **Import quotas into the USA and the EC may be reduced.** This has already happened in the USA and it may continue. In the EC where the Protocol is a permanent agreement, it is less likely — but pressure to reduce cane imports exists.

— **Alternative uses for cane may become more or less economic, and new ones may be developed.** Brazil has developed an extensive industry producing ethanol for car fuel. This may become less attractive if mineral oil prices continue low. Lysine, an amino acid valuable as an animal feed, can be produced on an industrial scale from sugar beet — and biotechnology may produce other possibilities. Some of these developments would produce additional markets for the beet or cane — either of which would help Third World countries.

Conclusions

The Panel noted the relative lack of pressure on the EC to change the CAP's arrangements for sugar, and also the possible developments in the sugar world over the next few years, most of which would work to the detriment of poor sugar-workers in the Third World. It also noted the historical responsibility of EC member countries for setting up Third World sugar industries in colonial times. Because of this, the Panel welcomed the permanent nature of the Sugar Protocol, which ensures continued high priced imports from some of the poorest people in the world.

The Panel received conflicting evidence on the effect of lower EC sugar exports. The immediate effect (perhaps a 10% increase in the world price) might be short-lived as new producers developed their products. On the other hand, if there were any small permanent price rise, it would help countries largely dependent on sugar exports. The Panel believes that the EC should not export any sugar — in other words that it should subsidise production only up to the level of its own consumption.

New uses for sugar (cane or beet) would also help Third World countries; new uses for beet would reduce the surpluses from the EC and the USA; and new uses for cane would provide new markets for Third World producers. Either might increase the world price. So the EC should use its research funds to develop new uses. There are dangers in this approach; the new uses could compete with existing Third World products, and the poor can suffer from expanding markets if they are driven off their land to make way for increased production. Care needs to be taken to avoid these dangers.

Finally, since the outlook for sugar is so precarious, urgent attention should be devoted to helping Third World producers to diversify. In most cases it is easier to call for diversification than to say how it should be done. But that should not deter us.

Box 5.A

THE EC's SUGAR SYSTEM

Every five years the EC sets production Quotas:

'A' Quota sugar is the amount the EC is expected to need for its own consumption.

'B' Quota sugar is an additional amount to cover unexpected shortfalls in production or increases in consumption.

'C' sugar is any sugar produced outside 'A' and 'B' Quotas. Unlike 'A' and 'B' Quota sugar, it does not receive any subsidies and has to be sold on the open market. In 1985 farmers received £6–£8/tonne for 'C' sugar, compared to about £27 for 'A' Quota sugar.

Box 5.A cont.

Price support

If 'A' or 'B' Quota sugar is sold on the world market, somebody has to pay the difference between the EC price and the world market price (currently around £20/tonne). To avoid paying all of this itself, the Commission has devised a system of Producer (or 'Co-responsibility') levies — which make the producers contribute towards the cost of selling the surplus on the free market.

Both 'A' and 'B' Quota sugar get price support — while 'C' sugar does not. The levy on 'A' Quota sugar is very low (2% of the intervention price), because none of it is supposed to be for export. But the levy on 'B' Quota sugar is 39.5%. The levies are paid 60% by growers and 40% by processors.

Source: The Hunger Crop, by Belinda Coote, Oxfam, 1987

Box 5.B

WORLD SUGAR

Main producers, exporters and importers, 1984

Million tonnes (raw value)

Cane producers	**Production**	**Exports**	
Brazil	9.3	3.0	
Cuba	7.8	7.0	
India	6.6	.3	
Australia	3.6	2.6	
Mexico	3.3	—	
Philippines	2.6	1.2	
S. Africa	2.3	.7	
Others	26.2		
Total cane	**61.7**		
Beet producers			**Imports** *(inc. cane)*
EC	13.3*	4.4	1.6
USSR	8.8	.2	5.5
USA	2.8	—	3.0
others	12.7		
Total beet	**37.5**		

*** includes 0.3 million tonnes of cane**

Source: Sugar Yearbook 1984, International Sugar Organisation

Box 5.C

THE SUGAR PROTOCOL

The Sugar Protocol of the Lomé Convention (see Box 1.B, page 13) allows cane sugar from 16 Third World countries to be sold in the EC free from the usual import restrictions, at the same price as EC sugar (though transport and insurance costs mean that the net price is less). There are similar arrangements for India.

Each country is allowed only a fixed amount (or quota). Currently the EC price is way above the world price (£26/tonne) so these countries get both a guaranteed market and a substantial subsidy. The exact size of the subsidy varies, because it depends on the difference between the world price and the EC price. In 1981–2 it totalled £105 million. The Protocol is a permanent agreement — it is not dependent on the continuation of the Lomé Convention.

Although 16 countries (plus India) benefit, three-quarters of the benefits go to five countries — Mauritius (38% of the total), Fiji, Guyana, Jamaica and Swaziland.

Members of the Protocol do not necessarily have to produce the sugar themselves, they are free to import it at the world price, and then simply re-export it to the EC. In its turn, the EC re-exports, because it is more than self-sufficient without the Protocol imports. The costs of transport, handling, insurance and waste amount to about 20% in the sugar trade — losses involved in these re-exports have been estimated at £22 million in 1981/2.

Some Protocol members pay their producers a price between the EC price and the world price — regardless of how much they produce. This encourages production beyond the quota — and thus lowers the world price.

Box 5.C cont.

SUGAR PROTOCOL QUOTAS

1981–2

	Tonnes	*% of total quota*
Mauritius	487,000	37.8
Fiji	163,600	12.7
Guyana	157,700	12.2
Jamaica	118,300	9.2
Swaziland	116,400	9.0
Trinidad & Tobago	69,000	5.4
Barbados	49,300	3.8
Belize	39,400	3.1
India	25,000	1.9
Malawi	20,000	1.6
St Kitts & Nevis	14,800	1.1
Madagascar	10,000	0.8
Tanzania	10,000	0.8
Zaire	4,957	0.4
Suriname	2,667	0.2
Uganda*	409	—
Kenya	93	—
Total	**1,288,826**	**100**

* quota abolished 1981

Sources: Lome III. The Courier, No. 89, Jan–Feb 1985, The European Community, Brussels
World Bank Development Report, 1986, p. 143, IBRD, Washington

Case-box 5

THE SUGAR-WORKER — JAMAICA

Nelsetta Johnson is one of 47,000 Jamaicans who earn a living from the island's sugar industry. Jamaica exports 118,300 tonnes to the EC under the Sugar Protocol of the Lomé Convention (see Box 5.C).

Nelsetta lives in a small wooden hut in the middle of the Frome Estate, one of Jamaica's state-owned sugar plantations that is managed by Tate & Lyle. The house is provided rent free by the Estate. It has neither water nor electricity. She shares its one room with her two daughters, her three year old son and her mother, now crippled after years of working in the

Photo: Belinda Coote

Scattering fertiliser on the cane fields.

Case-box 5 cont.

cane fields. Each morning she rises at dawn to prepare breakfast for her children and herself for work.

At 7 am the factory siren signals the beginning of the working day. Nelsetta is one of a gang of six women cultivators assigned to spreading fertiliser on the fields. Each woman has to provide her own plastic or wooden container for the fertiliser. It is brought to the fields in sacks, the containers are filled — and for eight hours a day the women trek back and forth across the ploughed fields, scattering the fertiliser as they go.

The women protect themselves from the burning sun and active ingredients by putting on layer after layer of clothing. They are not provided with any protective clothing by the management even though some brands of fertiliser that they have to use cause severe sores to their hands.

At mid-day they break for an hour. Frequently their lunch consists of only boiled yams or green bananas, sometimes flour and water dumplings.

As a cultivator Nelsetta is paid the minimum agricultural wage of £1.50 per day. But her take home pay is considerably less: for one 40 hour week in January 1986 she took home just £2.94 which means that she worked for seven pence an hour.

After nearly forty years of work in the cane fields Nelsetta's mother receives a pension from the company of just £2/week. This brings their weekly joint disposable income to £4.94.

The cost of living in Jamaica is high. A group of nutritionists calculated the basic food needs for a family of five for one week. Their shopping list to meet these requirements would have cost £22.50 in January 1986, more than four times the combined incomes of Nelsetta and her mother. Here is how long Nelsetta has to work to earn the price of basic foodstuffs — compared with an English sugar-beet worker on the minimum wage*:

	Nelsetta	UK worker
1 lb rice	3 hrs 20 mins	—
2 lb loaf of bread	—	12 mins
2 lbs potatoes	4 hrs 36 mins	5 mins
1 lb of spring greens	1 hr 23 mins	4 mins

Some of the work on the estate is better paid. As is so often the case it is the jobs which are normally assigned to women — weeding, planting, fertilising — which are paid at the lowest rates. This leads to special hardships in Jamaica where many households are headed by women.

Even with the advantage of Jamaica's high-priced exports to the EC, Nelsetta's life is one of plain hard labour and grinding poverty. If this ad-

Case-box 5 cont.

vantage were lost, her life, and those of her 47,000 co-workers would be even harder.

* based on an agricultural worker's minimum wage for 40 hour week, one adult and 3 children under 11 living in tied cottage. Granny (on pension) looks after children.

Sources: Belinda Coote, Oxfam.
Agricultural and Allied Workers National Trade Group.
Child Poverty Action Group.

6 CONCLUSIONS

The Panel recognised the EC's responsibilities in the world of trade. The EC forms the world's largest agricultural market and it is the largest trader in agricultural products. Clearly its first responsibility is to the people who live inside the Community — the 321 million whose interests it is set up to serve. But its responsibility does not rest there; in order to live up to its ideals, and those of the individual farmers, tax-payers and consumers whom it represents, it must treat its trading partners fairly — especially those who have little power and a lot to lose.

The Panel were not under the illusion that the CAP was the sole or even the dominant factor in world agricultural trade; they were aware of the importance of other exporting countries (the USA in particular), other importing countries (the USSR and its wheat imports, for instance), the influence of Trans- National Corporations (TNCs) and the fluctuating value of the dollar. The Panel's task was to concentrate on the CAP as one of the major factors in world agricultural trade — and one under strong and increasing pressure to change. In doing so, they did not forget the other factors.

In general, the Panel agreed that the CAP helps **farmers** (especially larger ones) inside the EC, but retards those in the Third World — while it works to the disadvantage of **consumers** inside the EC but helps food consumers in the Third World, at least in the short run. In the longer run, however, it is the Third World farmers on whom future Third World food production depends. And they are certainly harmed by CAP protectionism.

The Panel were also clear that — even leaving aside for a moment all moral, religious, humanitarian or ideological aspects — it is in the interests of the rich countries that the poor should get richer. As incomes increase, so does profitable trade.

In the light of the evidence presented to the Enquiry, then, the Panel call for action on five main points;

1 Reduce the surpluses

The effect on Third World countries of continuing policies that result in EC surpluses is in general damaging. The beneficial effect (cheaper food for poor people) could be obtained in other ways — for example through aid programmes. It is in the interest of the Third World (as well as of the EC) to reduce, eventually to zero, the EC's surpluses.

The Panel heard evidence on the merits of the different ways of reducing the surpluses — quotas, rationing fertilisers, leaving land fallow, alternative crops. From the Third World's point of view, the key point is whether a given method would lead to reduced demand for a Third World product, or to increased production of an alternative crop which competes with a Third World product. From this point of view, leaving land fallow and promoting alternative noncompeting crops such as trees would be best. Rationing fertilisers would do the same thing, but the Panel had doubts about its practicality.

The Panel argue that when decisions are made on the different methods of reducing the surpluses, the Community should take into account the effects on the Third World — and should instruct the Commission to bear this in mind when preparing proposals.

One way of reducing the surpluses is to make the CAP more responsive in some way to the world market. Some ways of doing this (for example a fixed instead of a variable import levy, which would lead to more fluctuating prices for both farmers and consumers in the EC) are probably not acceptable. But the Panel is clear that some way has to be found.

2 Build up Third World purchasing power

The best way to help the poor is to build up their purchasing power. The EC can help to do this two ways.

— **Offer the market to those who need it.** The EC's policy of increasing its self-sufficiency in all products is both expensive, and harmful to the Third World. Through the CAP, EC farmers can be paid for growing a wide variety of useful crops — or indeed for leaving their land fallow. In the Third World on the other hand, there are often no alternative money-making crops. Meanwhile it costs nearly £700 to produce a tonne of rapeseed oil in the EC and under £150 to produce a tonne of palm oil in Indonesia.

It would be better if the EC took more opportunities to offer markets to the Third World. In the first instance it should stop taking them away, as it does whenever it restricts or replaces imports from the Third World.

The Panel considered the use of a 'Social Clause' — a device to ensure that the market would go to those who most needed it. There are many practical difficulties involved, but it recommends that the EC consider how to devise a way of giving priority to poor countries, and to ones where the benefits of extra income are likely to be well spread.

— **Help the poor get maximum value from what they produce.** At present the CAP discourages Third World countries from processing their own agricultural products — by taxing processed imports more highly (see Table 1.1, page 9). They thus lose a considerable amount of income. The system suits the EC — it provides both money and jobs.

The Panel recognises the consequent difficulty of changing it — but it believes that in the long run it is in the interest of the EC that the Third World's purchasing power should be increased. Reducing tariffs on processed goods would be a prime way of achieving this.

The Panel heard evidence on the influence of the Trans-National Corporations (TNCs) on world agricultural trade, an influence which is not always helpful to poor people. It recommends that the EC take into account the policies and pressures of TNCs when changing the CAP.

3 Make the EC a dependable trading partner

The present arrangements mean that the EC is living at the edge of — or beyond — its budget. This results in rapid and not easily predictable changes (for example milk quotas). Such changes are likely to continue while the EC is on the brink of being unable to afford its own policies — and while the cost of its policies can be partly determined by world prices and fluctuating exchange rates. They are damaging to Third World countries, making it hard to plan or invest.

4 Find out who will be hurt

The most striking point to arise from the Enquiry is that almost any change will benefit some people in the Third World but harm others. Obviously there is no need to worry about the beneficiaries, but the Panel were most concerned about those who would lose from any change in the CAP.

The first step must be to identify those in the Third World who would lose (groups of people as well as countries). The EC has made one-off assessments in the past[25] — but what is needed is an assessment of each proposed change to the CAP as it comes up. This need not be as difficult as it looks at first; quite a lot of work (including computer modelling) has already been done in research centres and in government departments. Short-term consultants could be employed to report on the effect of any particular proposed change.

A simple (and small-scale) machinery in the Commission — consisting of existing staff and/or limited but adequate funds to commission outside work — would automatically commission a report whenever a change in the CAP was proposed. The report would not only identify Third World losers if the change were put into practice, but also suggest ways of modifying the change to soften its adverse effects. It would show the effects of a range of options, pointing out which were the most favourable (or least unfavourable) from the Third World's point of view. If changes to the proposals were not practicable, it would suggest appropriate ways of making good the harm done to Third World people.

The next step would be to make sure that effects on the Third World were known to the EC's decision-makers. So the report would be incorporated as a matter of course into the proposal sent to the Euro-Parliament and to the Council for decision.

Finally, the EC should accept that where changes in the CAP harm poor people in the Third World, the Community has a responsibility both to reduce such harm to a minimum, and to make good the damage in some way — through alternative trading arrangements or through aid. The Panel noted that the EC has already accepted the principle of making good the harm it caused in the case of the Thai cassava growers (see Box 2.D, page 26). It believes this principle should be extended to all cases where a change in the CAP causes harm to the poor.

5 International negotiations

The next 'round' of negotiations of the General Agreement on Tariffs and Trade (GATT — see Box 6.A) is to include agricultural trade for the first time. This offers both opportunities and dangers, both to the EC and the Third World. The opportunities are for the EC to find a way of safeguarding the interests of its own consumers and farmers without penalising its tax-payers or the Third World — and to develop itself as a dependable long-term trading partner with the Third World. The danger is that the power of the USA will force concessions on the EC — which will then be under pressure to claw back other concessions from the Third World. GATT negotiations go on for several years. The Panel believes that it is very important that the voice of poor people should be heard clearly throughout the negotiating period — and listened to by the rich countries.

Because the CAP affects different Third World countries in different ways, their voices are not always united. The present CAP arrangements suit some countries for some products, but work to the disadvantage of others. Not surprisingly, there is no single Third World view. But in the lengthy GATT negotiations the Panel recommends that the UK Government and the EC take account of the needs of the poor in each case.

The Panel also noted the proposal before the European Parliament to hold a World Food Conference to consider ways of creating a political and economic climate in which Third World countries could increase both food production and economic security. The proposal calls for liberalisation of agricultural trade and reduced agricultural protection. Its proposers feel that the GATT will either not tackle the problem adequately, or not with sufficient urgency.

The Panel believes that the evidence presented to it confirms the need for a proposal of this kind — and it strongly supports its aims.

6 The Lomé Convention and the least developed countries

The Panel noted the advantages to very poor countries of belonging to the Lomé Convention (see Box 1.B. page 13). But it was surprised that no less than ten of the thirty-six countries classified by the World Bank as the least developed are not members of the Convention*. Although the populations of some of these countries are very small, the total number of people who live in them is quite high (150 million) — most of whom are very poor.

The Commission believes that it is more appropriate to help non-Lome countries through other means — eg aid, food aid, STABEX (already extended to Bangladesh), or by special trade 'preferences'. Apart from Namibia when it wins independence, the list is closed. The Commission also points out that the existing members of the Lomé Convention are understandably not keen to extend (and thus perhaps dilute) the benefits they currently enjoy.

The Panel do not accept these arguments. While some of these countries might not wish to be associated with the EC, the Panel feel that that should be for them to decide (Haiti has already applied). The Community should surely extend the favoured terms of the Lomé Convention to any of the poorest countries in the world who wish to accept them.

* * *

In conclusion, the Panel believe that while changes to the CAP are bound to hurt some people in the Third World, forethought, care and political will can be used to minimise the harm. The task is complex, but it can be done — and common humanity demands that it should be done. Farmers in Europe and in the Third World share common ground in working the soil of one earth, and feeding the people of one world.

The solutions are not only technical; in the last decades of the Twentieth Century they more than ever involve economic power and political will. Farmers in the EC can play their part, not just by producing more food, but by generating the pressure for fairer trade and fairer food.

* Bangladesh, (pop. 98 million), Afghanistan (16m), Nepal (16m), Yemen (8m), Haiti (5m), Laos (4m), Democratic Yemen (2m) Bhutan (1m), Maldives (.2m), Samoa (.2m)

Least developed countries which are not members of the Lomé Convention

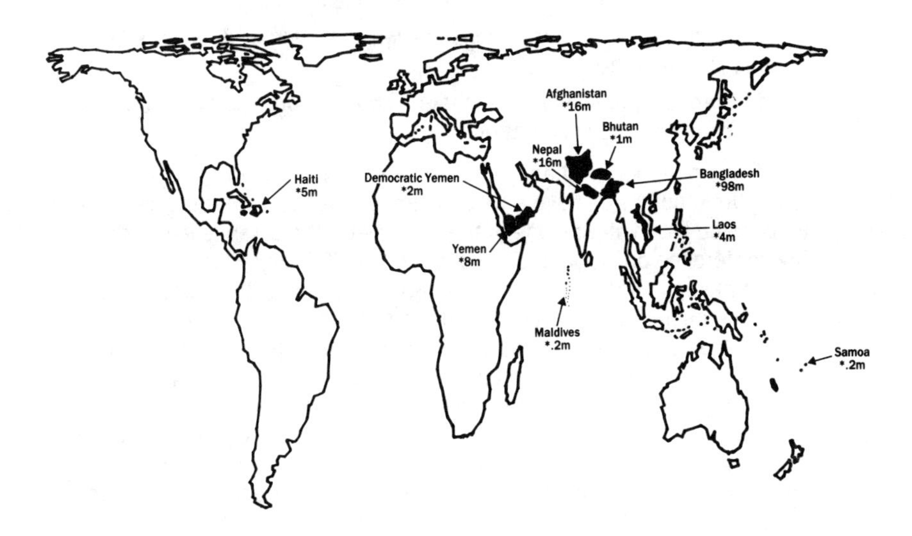

Box 6.A

The General Agreement on Tariffs and Trade (the GATT)

The GATT is the main international trade agreement. For nearly 40 years it has provided the framework and guidelines for much of world trade.

It began as a temporary arrangement arising out of the Havana Convention in 1948. It now has 92 members and 7 main regulations:

— Members agree to apply the tariff applied to their most privileged trading partner to all other Members (the 'most-favoured-nation' clause)

— customs duties should be the single method of protection

— Members agree to compensate supplying countries if they suffer from raised customs duties

— imports may not be controlled by quantitative restrictions — i.e. quotas. Exceptions are allowed for agriculture and for balance of payments difficulties

— emergency measures allow Members to raise duties or impose quotas if they are faced with a big increase in imports

— customs unions and free exchange zones are allowed, provided they do not harm others

— differences between Members should be settled through consultation and conciliation.

The Agreement was originally made by the industrialised countries, and Third World countries have always claimed that their interests were not considered — which was one of the reasons for the establishment of the UN Conference on Trade and Development (UNCTAD). But in practice the GATT makes two important concessions to Third World Members; they can avoid the basic rules in order to protect such areas of their economy as they judge "essential" — and the developed countries have agreed to abandon the principle of reciprocity by allowing some Third World products to be imported at low or nil rates of import tax.

The GATT has been changed surprisingly little; the only two major changes (or 'Rounds' as the negotiations are called) in its history have been the Kennedy Round of 1964–7 which handled the integration of the EC into the GATT system, and the Tokyo Round of 1973–9. Most agricultural products were excluded from both these negotiations, on the ground that they were "sensitive".

A third Round starts at the beginning of 1987 and is expected to last, like the other ones, for about four years. It will for the first time consider agricultural trade, a victory for the USA, Australia and other exporting

Box 6.A cont.

countries which view the EC's present arrangements as "unfair". The EC, however, has managed to get included in the negotiations "all direct and indirect subsidies", thus including methods used by the USA (see Box 3.B).

Sources: GATT: Agriculture, Rongead Infos, Vol. 86.2, Lyon. *Farm trade to come within GATT,* Financial Times, Sept. 22/86, London

Case-box 6

THE TOBACCO WORKER — BRAZIL

Brazil is the fourth largest tobacco grower in the world — 415,000 tonnes in 1984. It exports nearly half of it. About 40% of the EC's consumption is grown domestically in a protected market. Any change in the current level of protection (more or less) could affect imports from Brazil — and the 2½ million people, mostly small farmers and processing workers who depend directly on its cultivation there.

Tobacco is a very demanding crop, needing a lot of hard work — and the people who grow it, though not the poorest, are quite poor. The tobacco trade is their main source of income; their fragile standard of living depends on its continuation.

* * *

In Rio Grande do Sul in the south of Brazil, live Helmut (32) and his wife Maria (24), descendants of German immigrants. They have been tenants on an area of 13.3 hectares since they got married 7 years ago. They have two children of 5 and 1½.

They grow tobacco — but they have to give half of the harvest to the landlord. They are worried that this year what they get for the rest may not cover their inputs of fertilisers and pesticides. This year they are planting 30,000 tobacco seedlings (approximately 1½ hectares). They are also planting soya (they have to give 40% of that to the landlord).

If it wasn't for the landlord insisting, they say, they would not be planting tobacco, which is a very demanding crop in labour terms. Although they say they want to find another crop, tobacco is still their main source of income.

They also plant cassava, rice, beans, potatoes and greens for domestic use and they have 2 hectares under maize. Helmut works the land with oxen, as do most of his neighbours. They have chickens and

Photo: Frances Rubin

Helmut works the land with oxen.

Case-box 6 cont.

cows — but recently, to pay for medical care for their younger child, they have had to sell off 3 cows. They now have 6 left.

Work starts at 6 am and continues till 11.30 when they break for lunch. They start again around 1.30 pm and continue until they get things done. *"If we worked a set eight hours a day we'd go hungry"*. During the harvest there are times when they have to work around the clock, especially to keep the curing ovens going. November and December are the worst months, with May and June being a little easier. *"But in those months we have to prepare the wood for the furnaces. Farmers can't ever take holidays"*.

They buy in wheat flour, sugar and soap powder. Sometimes they buy coffee but mostly they drink Erva maté (tea) produced on the farm. They also make soap and have their own milk, eggs, cheese. Beer? *"we can't afford that!"*.

They cook with wood and sometimes gas. The monthly electricity bill is around £1.50 (electricity has been installed for about 8 years).

Clothes? *"That's difficult"*.

Leisure? Helmut used to play football, but he has not lately. They watch television in the evenings if there is time.

Source: Frances Rubin, Oxfam Recife.

NOTES AND REFERENCES

Introduction

1 Dietrich Busacker, *Satisfied livestock, degraded land and hungry people; feedstuff imports as a barrier to ecological society*, presentation to Workshops on Human Ecology, Edinburgh, May/85, BUKO, Hamburg

1 The Common Agricultural Policy

2 *World Development Report, 1986*, IBRD, Washington, 1986, p. 110.

3 European Community Commission, *Perspectives for the Common Agricultural Policy — the Green Paper of the Commission*, Brussels, July/85, p. II.

4 See Ulrich Koester and Malcolm D. Bale, *The Common Agricultural Policy of the European Community, a blessing or a curse for developing countries?*, pp. 4–7, World Bank Staff Working Paper No. 630, IBRD, Washington, Feb/84, for a more detailed account of the CAP's mechanisms.

5 Commission of the European Communities, Information Office, London

6 Commission of the European Communities, Information Office, London

7 Bureau of Agricultural Economics, *Agricultural Policies in the European Community*, Policy Monograph No. 2 (Canberra), 1985, quoted in *The CAP and its impact on the Third World*, ODI Briefing Paper, London, June/85.

8 Commission of the European Communities, Information Office, London

9 *World Development Report*, 1986, Table 6.5, IBRD, Washington, 1986, p. 119.

10 Economic and Social Committee of the European Communities, Official Journal No. C 330/19, Dec 20/85

11 *"We have suddenly become the villains of a play whose plot we don't understand. We are being made to feel guilty for doing our job too well. We are apparently producing food that nobody wants, snatching food from the mouths of famine victims to feed our livestock, and squandering tax-payers' money."* A farmer quoted in *Newsletter*, Farmers' Third World Network, No. 4, Winter/86, p. 7.

12 US Department of Agriculture

13 Unless otherwise specified, sources for the rest of this section are given on Charts 1 and 2

2 Animal Feeds

14 Dietrich Busacker, see Reference 1

3 Wheat

15 Commission of the European Communities, Information Office, London

16 Frans Andriessen, EC Commissioner for Agriculture, quoted in *The Guardian*, Nov 13/85 and Jan 8/86

Refugees' rations calculated at the rate of 500g per person per day — Oxfam Health Unit

17 Frans Andriessen, EC Commissioner for Agriculture, interview with NGO delegation, Oct 16/86

18 Alan Matthews, *Common Agricultural Policy and the Less Developed Countries*, Gill & MacMillan, Dublin, 1985, pp. 165–8

19 A. Sarris & Freebairn, *Endogenous Price Policies and International Wheat Prices*, American Journal of Agricultural Economics, 65:214–221, quoted in Matthews, op cit, p. 213

4 Food Aid

20 A useful account of Nigeria's dependency on imported wheat is given in Gunilla Andrae & Bjorn Beckman, *The Wheat Trap*, Zed Books, London, 1985

21 *Urgent Reform of European Food Aid needed says Overseas Development Minister*, Overseas Development, News Release, ODA, Oct 14/86

5 Sugar

22 International Commission for Solidarity among Sugar Workers, Toronto, The figure of 30 million is made up of 7 million workers and 23 million dependents. The Commission estimates that as many again are indirectly dependent on sugar — through processing, transport, services, etc.

23 *World Development Report, 1986,* Table 6.5, IBRD, Washington, 1986, p. 114.

24 Evidence to the Enquiry, Simon Harris, Table 5.

6 Conclusions

25 For example, *The Common Agricultural Policy and the EC's Trading Relations in the Agricultural Sector — effects on developing countries,* SEC (82), 1223, EEC Commission, July/82

APPENDIX

The Evidence

Besides the evidence which was presented to the Enquiry in person, a number of other submissions were made — some prepared specially for the Enquiry, some for other occasions. Brief summaries of the key points in the evidence submitted are given here. Inevitably they can only give a taste of the full submissions. Each summary has been agreed by the author. The full texts are available on request from Oxfam — or from the sources given if already published.

List of Witnesses

Geoffrey Bastin	Director, Oilseeds Research, Landell Mills Commodity Studies Ltd
Sir Richard Body MP	Chairman, House of Commons Select Committee on Agriculture
Brian Bolton	Agricultural and Allied Workers National Trade Group
Sipke Brouwer	Directorate General for Agriculture, Commission of the European Communities
John Cameron and **Belinda Coote**	School of Development Studies, University of East Anglia, Oxfam, Public Affairs Unit
Edwin Carrington	Secretary-General of the ACP Group, Brussels
Dr Edward Clay	Institute of Development Studies, Sussex University
Professor David Colman	Faculty of Economic and Social Studies Manchester University
John Ellis	Louis Dreyfus Trading Ltd
Susan George	Institute for Policy Studies, Washington
Professor David Harvey	Department of Agricultural Economics and Management, Reading University
Simon Harris	S & W Berisford PLC
Tony Jackson	Food Policy Adviser, Oxfam
John Jones	Farmer, Wales
Walter Kennes	Directorate General for Development, Commission of the European Communities

John Lingard and **Lionel Hubbard**	Department of Agricultural Economics, Newcastle University
Alan Matthews	Lecturer in Economics, Trinity College, Dublin
Sir Henry Plumb MEP	Chairman, European Democratic Group in the European Parliament
Brian Rutherford	United Kingdom Agricultural Supply Trade Association Ltd
Paul Rynsard	The Green Party, Agricultural Working Group
Dr David Seddon	Overseas Development Group, University of East Anglia
F. E. Shields	National Federation of Young Farmers' Clubs
Huub Venne	Free University of Amsterdam
Nick Viney	Dairy and Sheep farmer, Dorset

* * *

Geoffrey Bastin
Director, Oilseeds Research, Landell Mills Commodity Studies Ltd

1 The oilseed market is oversupplied — so a rational response would be for countries to reduce their production. But opportunities for alternative production are limited in many oilseed-producing regions — especially in Third World countries.

2 So there is pressure to continue production — often with high levels of support and protection. This pressure will be hard to resist.

3 Over-supply is partly due to the success of the EC's policy of encouraging oilseeds production in Third World countries.

4 From Third World countries' point of view, palm-oil is a profitable and reliable crop, making use of land unsuitable for other crops; it also provides a basic foodstuff, and one which must be processed locally, thus giving a base for industrialisation.

5 The EC is self-sufficient in:

Oilseed	15%
Oils	20%
Meal	5%

6 The Commission is unable to accept the success of its own external development programmes — and thus continues to subsidise domestic production of rape-seed. So it is a major exporter of rape-seed oil, which competes head-on with the palm-oil from the Third World whose production it has encouraged.

7 In addition, the EC's policy of subsidising rape-seed contributes to the surplus, which reduces the price for Third World exporters.

8 **Estimated production cost of veg oil, 1985**

Indonesia	palm	$185/tonne
Malaysia	palm	$240
USA	soybean	$463
Canada	rape	$648
EEC	rape	$900

9 These price differences put a heavy burden on the CAP. In 1985 oilseeds accounted for only 2% of agricultural production in the EC, but 10% of total budget allocation (2bn ECU). The EC also spends $150m on research designed to increase production in Third World countries.

10 So the EC is a major consumer of oilseeds. Any policy which reduces import demand (by self-sufficiency or protection) hurts the producing countries.

11 **Tariff barriers** are significant:

Soyabeans	0%
Soymeal	7%
Crude soyoil	10%
Refined soyoil	15%
Margarine	25%

Actual effective rates are higher.

12 **Spain** limits domestic consumption of soyabean oil and so it is a major exporter. With Spain now a member, the EC is the world's largest exporter of soyabean oil; ironically much of the raw materials for these exports are imported from Brazil where there is unused processing capacity.

Full paper, *The CAP and the Oilseeds Economies of Developing Countries*, available from: Landell Mills Commodities Studies, 50/51 Well Street, London W1P 3FD

* * *

Sir Richard Body MP

Chairman, House of Commons Select Committee on Agriculture

1 Britain should unilaterally declare a free trade policy. This would enable Third World countries to regain the chance to supply UK consumers with food (and give UK consumers the freedom to buy it). It would also extricate the UK from the process of dumping surplus food on the world market in competition with others who cannot afford to use their taxpayers' money in subsidies.

2 Some people would get hurt by this policy, but there would be major gains to several million people — and it would set an example for other countries to follow.

3 Britain set up the systems which produce the crops we are now not buying (transporting slaves and others to places which can only support themselves by growing export crops) — so we have an obligation to continue to buy those crops from them at fair and reasonable prices.

4 Although most of this is history, we are still benefiting from those early producers; cheap food (especially sugar, for example) enabled us to devote more money to our own economic development, and we are still enjoying the fruits of historical extra income.

5 The EC should also cut its exports of sugar — the surpluses undermine the ability of Third World producers to sell on the world market at a cost above that of production.

6 Third World sugar producers have so far resisted the blandishments of the USSR and Cuba; let us not make it difficult for them to go on doing so.

Evidence based on *Farming in the Clouds*, Chapter 8, 'The poorest victims', by Richard Body, Temple Smith, London, 1984, £3.50

* * *

Brian Bolton
Agricultural and Allied Workers National Trade Group

1 Neither the EC (which includes bureaucrats and agro-industries as well as farmers and consumers) nor the Third World should be considered as uniform bodies.

2 Agri-business Trans National Corporations (TNCs) have in many ways a more powerful effect than the CAP's effects through tariffs and quotas, or through depressing prices.

3 For most products one, two, or sometimes three TNCs regulate the conditions of trade. Manufacturers of foods and agro-chemicals are also very powerful. Their accountability is close to non-existent.

4 Unless we recognise the power and influence of the TNCs, virtually all of the well-meaning proposals that can come out of this debate become but empty rhetoric.

5 Third World countries are too vulnerable to take on TNCs when their interests are damaged; the rich countries are unwilling to do so. Most of the International Commodity Agreements are in disarray; the rich countries resist the establishment of commodity funds proposed by UNCTAD.

6 In the Third World subsistence farmers are the majority; the CAP affects mainly commercial farmers who are a minority.

7 Huge acreages in the Third World are farmed under direct contract to the TNCs. This is in contrast to the EC where there is only a limited amount of contract farming.

8 TNCs have a big advantage over Third World countries in their access to information (e.g. on differential prices, shortages, gluts, transport facilities, etc). The EC has the resources to make this kind of information available to Third World countries if it wished.

9 Farmworkers in the EC do not benefit much from the CAP; the Commission's Directives do not mention farmworkers, but assume that the wealth created by the CAP will 'trickle down' to the farmworkers. This does not happen.

* * *

Sipke Brouwer
DGVI (Agriculture), Commission of the European Communities

1 The CAP needs reform for reasons of production trends and budgetary costs. This reform is being carried out on the basis of art. 39 of the Treaty of Rome. It aims principally for a more market-oriented agricultural sector, with sufficient perspective for income and employment for its more than 12 million farmers.

2 The EEC has the largest market and agricultural sector of the world. As to agricultural products, the EEC, contrary to other industrial nations, imports twice as much as it exports. Nearly 50% of its agricultural imports originate from the developing world, so it has a big responsibility.

3 While reforming the CAP, the EEC does not intend to change this import situation. On the contrary through the Lomé III agreement and the new Mediterranean agreements, the EEC further guarantees the continuation and the development of traditional and new agricultural exports of the countries concerned towards the EEC market.

4 Why? Because the developing countries are predominantly agricultural economies, most of the time with a heavy debt burden. The Community must continue to buy their exports and provide for the purchasing power of these countries.

5 Such trade is necessary to enable countries to develop their various natural resources in an optimal way, and to use comparative advantages they have. Self-reliance in every product is not an optimal solution.

6 Hunger in the world has not been caused by the CAP, nor will the reform of the CAP end hunger in the world.

7 However, the CAP has proved very efficient in increasing agricultural products. It could serve as an example for a valid policy to develop the agricultural sector in Third World countries.

8 Naturally this can only be undertaken by national governments who could envisage for example an internal price policy that stimulates rural development through a reasonable price offered to the local farmers for

their products. Necessary imports at the present low world market price would be adopted at that internal level. The government could then invest the difference in priority policy sectors of the economy, such as rural development.

9 As to food aid and more generally the food supply of the Third World, the Commission envisages the following proposals:

a) improved procedures for food aid (including — triangular — under discussion in the Commission at present),

b) preferential supplies of food for Third World countries importing food.

* * *

John Cameron
School of Development Studies, University of East Anglia
Belinda Coote
Public Affairs Unit, Oxfam

1 It is impossible to make adjustments to sugar policies under the CAP which benefit all cane producers at limited cost to EC beet producers and no net cost to taxpayers and consumers in the EC.

2 All major sugar producers sell some sugar on the world market — so all countries (including those in the EC) suffer from low prices and instability.

3 It is difficult to allocate blame internationally for low prices — but the EC is accused of greater blame because it is a late-comer, because it exports a lot at heavily subsidised prices, and because it is said to have obstructed an effective International Sugar Agreement. So in the world market the EC is the international villain.

4 But with the Sugar Protocol the EC is the angel; the 16 ACP countries in the Protocol get good prices — and UK consumers pay more than they otherwise would. The EC is paying for imported sugar which it could easily produce itself, or buy more cheaply on the world market.

6 Protocol prices are linked to EC ones — so if EC farmers are paid lower prices, so would Third World ones be. In this case therefore, the interests of EC sugar farmers and Third World ones coincide.

Proposals

a) 5% annual reductions in sugar quotas for EC beet producers for 5 years. Farmers would be paid to use the land for other purposes (crops or leisure)

b) Maintenance of a price structure giving adequate prices to smaller EC beet producers — and thus adequate incomes to cane producers

c) Extend the Sugar Protocol by ½-million tonnes — to be supplied by countries where the benefits arising from the extra income would be widely spread

d) Extend aid to help Third World countries achieve greater food self-sufficiency and to develop more industrial local uses for cane.

Full paper, *Who should grow sugar for British consumers — and what price should they pay?*. Evidence to the Oxfam Hearing on the Effect of the Common Agricultural Policy on the Third World, Bury St. Edmunds. June 4/86, available from: Publications Officer, School of Development Studies, University of East Anglia, Norwich NR4 7TJ

* * *

Edwin Carrington
Secretary-General of the ACP Group, Brussels

1 There is a need in the Lomé Convention for consistent policies — for example the EC blocks sugar imports while financing a refinery in W. Africa.

2 The CAP is not designed for the benefit of Third World countries — therefore the most we can expect in relation to our interest is 'adjustment at the edges.'

3 This is critical; 27 of the 36 least developed countries are not in Lomé, and they depend primarily on agriculture. Given their small size, their interests could be accommodated at little cost to the EC.

* * *

Dr Edward Clay
Institute of Development Studies, Sussex University

1 The historical record for the effects of Food Aid on development and agriculture is inconclusive.

2 Africa is a particularly difficult case — because of limited scope for industrial growth, EC surpluses being different from local staple foods, and bad management.

3 The EC lacks 'staying power' in its Food Aid policies — it tends to divert its energies into topical initiatives. It would do better to go for long term consistency.

4 The EC needs to improve its capability to respond flexibly and rapidly to emergencies; food can be delivered in 6 weeks — the EC average is 24 weeks. EC managers should be made responsible and accountable — and able to take rapid decisions. If food aid is not delivered promptly, it can do more harm than good — by clogging up the ports and flooding the market with free food just when local farmers are wanting to sell.

5 The EC should keep stocks available for emergency distribution.

6 The EC's tendering system is very inadequate — if saving lives were the priority, it would not be organised in this way.

7 There should be more flexibility — including 'triangular' trade — buying food from a nearby country for use in the disaster area.

8 Food Aid should be linked to a food strategy.

* * *

Professor David Colman
Faculty of Economic and Social Studies, Manchester University

1 EC protection against imports reduces the value of Third World export earnings — it is 'a major constraint' on them.

2 EC exported surpluses have kept down prices for Third World consumers — one estimate calculates that the number of 'hungry' people in the Third World is reduced by 25% as a result of the CAP's protection.

3 Trade liberalisation would benefit consumers more than producers inside the EC — but outside it would be the other way round. The CAP operates to the disadvantage of EC consumers, but to the advantage of Third World ones; it benefits EC farmers, but harms the interests of those in other countries.

4 The relatively few opportunities which the CAP has created for Third World countries to increase their exports to the EC (e.g. cassava, etc) are insignificant when matched against the negative trade impact of the CAP's overall import restrictions.

5 Third World countries would gain relatively little from CAP liberalisation — one estimate is that world trade would increase by 39%, but would only increase the gross output of Third World countries by 2%. Even complete dismantlement of the protection is not projected to have a dramatic impact on total output.

6 But reduced trade barriers would have a very large impact when compared with that of aid, and there would be benefits in reduced instability.

Full paper published as: *EEC Agriculture in conflict with trade and development*, Manchester Papers on Development, Issue No. 10,Nov/84 — available from Publications, Dept. of Admin Studies, University of Manchester, Precinct Centre, Oxford Road, Manchester M13 9QS

* * *

John Ellis
Louis Dreyfus Trading Ltd

1 Changes in the CAP's cereals policy should not have too much direct effect on Third World countries:

— the CAP is not the main determinant of world cereals prices; the two main ones are **USSR harvests** and the **US price** (which is currently set to fall)

— so any short-term effect would fairly soon be eclipsed by these.

2 The short-term effect would be to raise world prices — which would hurt the small number of exporters and benefit the countries which import.

3 But **the method** by which the EC reduces its production may affect Third World countries, especially if home-grown cereals replace animal feed imports or if land now used for growing cereals is switched to growing things currently imported from Third World countries — e.g. oilseeds. New uses for cereals, though unlikely, could also affect Third World countries if the new products replaced current imports from the Third World.

4 Even if the EC manages to reduce cereals production, it still has the problem of the surpluses — which would be very expensive and disruptive to dispose of on the world market.

5 An alternative — although an expensive one — would be to dispose of them through non-commercial channels to relieve the famine in Africa.

6 The current outlook is that low prices will be with us for some time to come — barring another and more damaging Chernobyl-type incident, combined with crop failure in either the USA, India or Europe.

* * *

Table 1

World Wheat Trade *(M Tonnes)*

	1975/76	1983/84	1984/85	1985/86
Imports				
Western Europe	7.7	5.5	3.2	3.7
Eastern Europe	4.0	3.5	3.0	3.1
USSR	10.2	20.6	28.3	17.5
North America	2.0	3.7	3.4	3.1
South America	6.8	8.8	9.0	7.3
Near East	3.3	10.8	11.1	10.2
Far East	22.7	28.0	24.9	23.3
Africa	9.5	18.5	19.8	18.3
Others	0.3	0.9	1.1	1.5
TOTAL	**66.5**	**100.3**	**103.9**	**88.0**
Developed Countries	13.9	12.1	10.5	10.8
Developing Countries	34.0	51.6	52.6	47.2
Centrally-Planned	18.6	36.6	40.8	30.0

Table 1 cont.

Exports				
Argentina	3.1	9.6	8.0	6.5
Australia	8.1	11.6	15.1	15.5
Canada	12.1	21.2	19.1	17.5
EEC	7.7	14.9	17.4	15.5
USA	31.5	38.3	38.0	26.0
USSR	1.0	0.5	0.8	1.0
Others	3.0	4.1	5.6	6.0
TOTAL	**66.5**	**100.3**	**103.9**	**88.0**

Source: International Wheat Council

Table 2

Shipments of Grain Under the Food Aid Convention *('000 T)*

Donor Member	1981/82	1982/83	1983/84	1984/85
Argentina	5	33	30	51
Australia	535	325	565	541
Austria	20	33	22	20
Canada	600	828	815	947
EEC	1,833	1,622	2,046	2,738
Finland	—	20	40	20
Japan	309	310	311	363
Norway	27	23	12	43
Spain	22	16	21	25
Sweden	39	37	46	40
Switzerland	22	31	31	62
USA	4,791	5,862	6,453	6,976
TOTAL	**8,204**	**9,139**	**10,392**	**11,825**

Table 3

Recipients of Grain Under the Food Aid Convention, 1984/85 *('000 T)*

	Wheat	Rice	Coarse Grains
Central America	424	49	293
South America	181	57	40
Near East	105	13	—
Far East	2,364	243	108
Africa	4,306	445	1,644
Others	5	2	4
TOTAL	**7,385**	**809**	**2,090**

Source: I.W.C.

Table 4

EEC Wheat Production, Imports, Exports, Usage and Carryover Stocks *('000 T)*

	1975/76	1983/84	1984/85	1985/86
Production	37,696	59,236	76,400	67,000
Imports	7,124	3,123	2,700	3,100
Domestic Usage	37,504	49,328	53,200	52,000
Exports	9,481	15,583	17,200	15,500
Exports	9,481	15,583	17,200	15,500
Carryover Stocks	8,256	8,207	17,000	19,600

* * *

Susan George
Institute for Policy Studies, Washington

1 The effect of protectionist policies has surely been to dispossess Third World farmers — in 1984 the EC put in $17,000 million of subsidies and the USA $19,000 million. How can the Third World compete with subsidies on that scale?

2 There is far more emphasis on cash crops rather than subsistence ones; 30% of the EC's aid budget is still devoted to cash crops. And the Lomé Convention is mainly about cash crops.

3 The CAP does not contribute what Third World countries need:
— stable or guaranteed prices
— higher prices to help producers.

4 We could learn from the example of the Marshall Plan. In the post-War period the USA needed a valid trading partner. That is what the EC needs now if it is not be left behind.

5 Research shows that when a Third World country's income rises, so do its imports — especially of food and paper. So it is in the interests of the rich countries that Third World countries get richer.

* * *

Simon Harris
Group Economist and EEC Advisor — S & W Berisford PLC

1 Sugar is in structural surplus and likely to remain so. World market prices represent only residual values and are not representative of production costs.

2 This is because all governments protect their sugar industries — including Third World governments (for employment, food security and foreign exchange reasons). Production is thus insulated from the world market, as is consumption in most countries. Very few countries, whether developed or developing, allow their consumers to buy sugar at 'free' world prices.

3 The long-term future for sugar depends on the development of new uses for sugar, and encouraging some countries to diversify out of sugar.

4 The sugar cycle (higher prices recurring every 7 years or so) is likely to become longer, with lower peaks. This will deprive the weaker industries of funds to re-invest and modernise. There will thus be continued concentration in the world's sugar industry.

5 Non-sugar sweeteners are a threat, but contained in the EC by protection and in Third World countries by problems of technology.

6 Many Third World countries are members of Special Arrangements which protect their exports. Those with a high proportion of unprotected exports are Fiji, Nicaragua, the Philippines, Swaziland, Taiwan, Thailand and Zimbabwe. These are particularly vulnerable to falling world demand and prices. Developing country exporters particularly interested in continuing to expand their exports include Brazil, Cuba and Thailand.

7 If the EC ceased to export sugar, prices would go up by about 10% — but only for a while; other producers would soon fill the gap. So the benefit to Third World countries exporting sugar would be fleeting.

8 There is a continuing trend towards white sugar. This will increase the vulnerability of sugar exporting countries with small scale industries — which cannot justify the installation of white sugar facilities.

9 Sugar is not high on the EC's reform of the CAP list;
— there are not the same intractable surpluses as with (say) cereal or dairy products;
— it is self-financing (in terms of the Community Budget) and self-adjusting to the state of the world market;
— there are not the same management difficulties.

* * *

Table 1

Expansion in sugar exports from 7 countries *(million tonnes raw value)*

	1971	1981	Change	
Argentina	0.121	0.709	0.588	+ 86%
Australia	1.779	2.982	1.203	+ 68%
Brazil	1.230	2.670	1.440	+117%
Cuba	5.511	7.071	1.560	+ 28%
Swaziland	0.159	0.345	0.186	+117%
Thailand	0.145	1.155	1.010	+697%
EEC	1.288	5.344	4.056	+315%
	10.233	20.276	10.043	+ 98%

Table 2

Growth in cane and beet sugar yields per hectare over the last fifteen years for selected countries *(tonnes raw sugar/ha)*

	Three year average yield 1967/68–1969/70	Three year average yield 1982/83–1984/85	Change	
			tonnes	%
North America				
USA—cane	6.3	7.2	+0.9	+ 14.3
USA—beet	5.2	5.9	+0.7	+ 13.5
Mexico—cane	4.5	6.5	+2.0	+ 44.4
USA—Hawaii	24.0	26.1	+2.1	+ 8.8
Caribbean (all cane)				
Cuba	5.2	5.2	0.0	0.0
DR	5.4	6.1	+0.7	+ 13.0
Jamaica	6.4	4.7	−1.7	− 26.6
Trinidad & Tobago	6.1	2.8	−3.3	− 54.1
Central America (all cane)				
Belize	4.3	4.6	+0.3	+ 7.0
El Salvador	5.2	7.3	+2.1	+ 40.4
Guatemala	3.6	8.1	+4.5	+125.0
Nicaragua	3.6	6.3	+2.7	+ 75.0
South America (all cane)				
Argentina	4.3	6.5	+2.2	+ 51.2
Brazil	5.7	4.7	−1.0	− 17.5
Colombia	10.3	12.5	+2.2	+ 21.4
Guyana	6.6	5.4	−1.2	− 18.2
Peru	11.5	13.7	2.2	+ 19.1

Table 2 (continued)

Growth in cane and beet sugar yields per hectare over the last fifteen years for selected countries *(tonnes raw sugar/ha)*

North Africa				
Egypt—cane	5.4	8.8	+3.4	+ 63.0
Morocco—beet	5.2	6.7	+1.5	+ 28.8
Southern Africa (all cane)				
Kenya	2.7	9.7	+7.0	+259.3
Mauritius	7.0	8.3	+1.3	+ 18.6
South Africa	7.5	7.9	+0.4	+ 5.3
Swaziland	11.8	12.1	+0.3	+ 2.5
Zimbabwe	6.1	13.4	7.3	+119.7
Asia (all cane)				
China	2.4	4.8	+2.4	+100.0
India	3.6	2.5	−1.1	− 30.6
Indonesia	5.4	6.4	+0.8	+ 16.7
Philippines	4.8	5.6	+0.8	+ 16.7
Thailand	1.8	3.9	+2.1	116.7
Oceania (all cane)				
Australia	10.6	11.1	+0.5	+ 4.7
Fiji	7.0	6.9	−0.1	− 1.4
EEC (all beet)				
France	5.4	8.6	+3.2	+ 59.3
W. Germany	6.1	7.5	+1.4	+ 23.0
Italy	4.3	6.1	+1.8	+ 41.9
UK	5.2	7.6	+2.4	+ 46.2
EEC Average	5.2	7.6	+2.4	+ 46.2
Eastern Europe (all beet)				
E. Germany	2.7	2.7	0.0	0.0
Czechoslovakia	4.3	4.0	−0.3	− 7.0
Poland	4.3	4.2	+0.1	+ 2.3
Yugoslavia	4.3	5.8	+1.5	+ 34.9
USSR	2.4	2.4	0.0	0.0
Middle East (all beet)				
Iran	2.4	3.8	+1.4	+ 58.3

Note: The raw data for the years 1967/68, 1968/69 and 1969/70 was quoted by the USDA in short tons rounded to the nearest whole number. The numbers shown above are the result of the averaging and conversion of the original USDA data and hence, despite their quotation in the table to a single figure of decimals, cannot have any greater accuracy than the original data.

Sources: USDA (1969, 1970, 1975, 1981 and 1985).

Table 3

The range in world sugar total production costs.[a] 1979/80–1982/83 *(weighted world average = 100)*

Country	sugar plant	index value
Japan	cane	229
Japan	beet	170
Trinidad[c]	cane	158
Italy[b]	beet	155
Poland	beet	141
Guyana[c]	cane	118
Barbados[c]	cane	113
US	beet	111
US mainland	cane	107
UK[b]	beet	102
Jamaica[c]	cane	97
W. Germany[b]	beet	92
Belize[c]	cane	91
India	cane	90
Thailand	cane	
Argentina	cane	79
Brazil N/NE)	cane	
Philippines	cane	75
Mauritius[c]	cane	
France[b]	beet	73
Cuba	cane	71
Dominican Republic	cane	69
St. Kitts[c]	cane	66
Australia	cane	62
Fiji[c]	cane	60
Brazil (Central/South)	cane	57
South Africa	cane	

Note: a) Average 1979/80 — 1982/83 exchange rates and base years for data; constant prices; ex-mill costs in terms of raw sugar (96 pol). Costs derived by a synthetic cost engineering approach.

b) EC Member State (c) ACP State with an EC sugar quota.

Source: Derived from Congressional Research Service (1985, Chapter III).

* * *

Professor David Harvey
Department of Agricultural Economics and Management, Reading University

1 The co-responsibility levies charged on producers for sugar on C quota production are supposed in theory to reduce the effective price of this sugar

to the world price. In practice they never do — but the effects of over-production are much more limited than for products where support is unlimited.

2 Protection practised by other developed countries contributes to instability — so estimates of the EC's contribution to world instability are contentious.

3 Some research suggests that some of the effects of instability find their way back into the EC.

4 The probability is that internal pressures on the CAP will result in significant limitation to the expansion of EC products to the rest of the world at subsidised prices. This will help Third World food exporters. Thus domestic changes will limit further damage.

5 The major exception to this is likely to be import levies on cereal substitutes — e.g. cassava.

Proposal

The EC should re-direct funds currently used to subsidise domestic (EC) production to subsidise instead Third World food production, thereby raising world prices, and allowing production in both the EC and the Third World to respond accordingly. *"Of course, this requires a degree of international co-operation which seems beyond the bounds of possibility."*

* * *

Tony Jackson

Food Policy Adviser, Oxfam

1 Only very small amounts of dairy food aid are needed for emergency use in nutritional centres and populations in disasters.

2 The use of dairy food aid in non-disaster times is very dubious. It is difficult to help Third World countries increase their own food production by giving free milk powder. Operation Flood in India *may* be an exception — but it is controversial, and probably not replicable elsewhere.

3 If we continue with food aid as it is being allocated and managed at present, the forecast that Africa will continue to need to import food for the next 15 years may become self-fulfilling.

4 The cost of dairy food aid is very high (⅔ of all EC food aid).

1 **Conclusions:**

i) Dairy food aid is both costly to provide and hard to use productively in the Third World. The problems of over-production in the EC cannot be solved by dairy food aid

ii) Instead of non-emergency food aid, we should encourage Third World governments to improve their agricultural policies — at present they can come as supplicants to the EC — which is very sympathetic because it is so keen to get rid of the surplus.

* * *

John Jones
Farmer, Wales

1 Quotas should be allocated on the basis of the number of full-time agricultural workers on a farm.

2 This would help keep rural employment up, and would discourage both expansion (farmers would have to take on more staff) and contraction (they would lose quota).

* * *

Walter Kennes
Directorate General for Development, Commission of the European Communities

1 Markets for food products are peculiar because people cannot easily adjust the amount of food they eat. Therefore, shifts in supply have relatively large effects on prices. Such price changes are harmful especially for poor consumers and small producers. These basic facts have led governments all over the world to intervene in food markets in a variety of ways.

2 There are many sources of instability in world market prices for food and agricultural commodities apart from the export policies of surplus producers. Importers that have sufficient foreign exchange reserves may disregard the prices they have to pay and thus amplify price movements. For example the massive imports of cereals by the Soviet Union had a major effect on world market prices (e.g. fertilisers) and exchange rates contribute to price fluctuations. Not much is known on the relative magnitude of the effects of different sources of instability.

3 A recent analysis (carried out at IIASA, Vienna) has estimated the likely effects on the world food situation of dismantling agricultural protection. The results indicate that total abolition of agricultural protection leads to some important shifts in agricultural trade and production, but has a relatively small effect on world hunger. The basic explanation is that hunger is caused by poverty which can only be addressed by increasing the resources (in the widest sense) of poor people. Therefore poverty-oriented policies and programmes remain vital to alleviate hunger.

4 According to various studies the effects of the CAP on developing countries are on the whole rather small even though for specific countries, commodities or periods they may be more important. Net importers of food

tend to benefit whereas net exporters tend to lose out. With a few exceptions (e.g. Zimbabwe) sub-Saharan Africa is a net importer of food.

5 As regards CAP reform the EC Commission intends to allow properly for the Community's trade relations with developing countries. This also applies in the context of the GATT negotiations. The EC remains by far the largest market for Third World agricultural exports and therefore has a special responsibility in this respect. For a group of countries the EC has set up a system (STABEX) to compensate for agricultural export earnings shortfalls, regardless the source.

6 Over the past few years the EC has reoriented its development cooperation (using a variety of instruments) towards strengthening the food system. Where possible this is done in the context of a food strategy.

* * *

John Lingard and Lionel Hubbard
Department of Agricultural Economics, Newcastle University

1 The CAP effectively insulates EC farming from world market trends.

2 Exports force price instability onto world markets — to the detriment of Third World countries which are heavily dependent on agricultural markets

3 Annual world cereals production and prices have varied much more in the last two decades than previously. The reasons are unclear — but a continued high variation in production and prices is likely.

4 This increased price variability had coincided with increased and variable EC cereals exports.

5 Research suggests that the CAP is responsible for depressed prices on world markets and over half the increased variability.

6 Assured grain imports are important to Third World countries, already trying to cope with very variable production (weather, disease, etc) themselves. The combination of local and world variability make planning for food security impossible. This applies to both imports and exports (it adds to revenue instability of exports, and increases risk).

7 The CAP is therefore an adverse influence on the food security of Third World countries and EC levels of stocks and prices should be made more sensitive to world market conditions.

8 Pressure from Third World countries to reform the CAP is unlikely to have much influence in the EC, but it may be possible to set up a continuing Food Aid programme to meet grain production shortfalls which exceeded, say 6%. Food Aid could be used to tackle instability in domestic production in Third World countries.

9 An alternative suggestion for reform is for a constant tariff to replace the variable levies. This would dampen world fluctuations and help Third World countries — but it would mean that EC farmers could not depend on a guaranteed minimum price, and EC consumers would find food prices more variable. There would thus be a direct trade-off between Third World and EC interests.

* * *

Table 1

World Wheat and Coarse Grains Trade, 1974/1975 to 1984/1985 *(totals in millions of metric tonnes)*

	1974/1975	1979/1980	Estimated 1984/1985
Total world production	**976.8**	**1,164.7**	**1,295.1**
United States (as % of total)	20.4%	25.5%	23.4%
Other major exporters*	9.8	7.9	8.7
Western Europe	14.5	12.6	14.6
Soviet Union	18.8	14.7	12.3
Eastern Europe	9.3	7.8	8.4
People's Republic of China	10.0	12.5	13.9
Others	17.1	18.9	18.8
Total world consumption	**987.0**	**1,184.9**	**1,273.9**
United States (as % of total)	14.2%	15.4%	14.6%
Western Europe	15.8	13.8	12.5
Soviet Union	19.5	18.1	16.4
People's Republic of China	10.5	13.2	14.9
Others	29.9	39.4	41.4
Cumulative surplus stocks	**115.1**	**172.0**	**190.6**
United States (as % of total)	23.5%	55.2%	43.9%
Foreign	76.5	55.2	56.0
Total Trade	**126.7**	**186.8**	**206.0**
Exports			
United States (as % of total)	48.9%	58.2%	49.9%
Other major exporters*	34.4	29.7	30.6
Western Europe	10.0	8.9	13.9
Soviet Union	3.9	0.3	0.5
Others	3.4	2.8	5.0
Imports:			
Western Europe (as % of total)	25.7%	16.4%	7.3%
Soviet Union	4.1	16.3	23.8
Japan	14.5	13.1	13.1

Table 1 cont.

Eastern Europe	8.7	9.4	3.7
People's Republic of China	46.9	5.8	5.0
Others	46.9	39.0	47.1

* Other major exporters are Argentina, Australia, Canada, South Africa, and Thailand.

Source: U.S. Department of Agriculture.

Alan Matthews
Lecturer in Economics, Trinity College, Dublin

1 The CAP has a complex set of trade effects which can be positive, negative, or both, depending on the Third World countries involved.

2 In the USA the Loan Rate (equivalent to the EC's intervention price) is set to fall dramatically (e.g. 25% for wheat and maize) — so the world price for cereals and other food commodities will fall too. Also, under the 1985 Food Security Act, stocks can be sold off — so there will be further price falls. Since the CAP's reform would lead to higher world prices, this is a good time to reform (from Third World importers' point of view).

3 Instability; if there is a global production shortfall, someone, somewhere must respond with higher prices and lower consumption. Since the EC is insulated from the world market, all this falls on the rest of world. Price stability in the EC is at the expense of greater instability outside it.

4 The CAP increases some world prices — especially animal-feed, for example; CAP reform would lower these prices, and eliminate markets.

5 The CAP's reform would erode the value of concessions to ACP countries under the Lomé Convention, and would lead to a fall in food aid.

6 The EC's agricultural protection makes employment creation in Third World agriculture harder — so there is a need to create more jobs in non-agricultural sectors. There will thus be increasing pressure on the EC from Third World countries to accept more manufactured products. But meanwhile the EC is more likely to protect itself against such imports — if only because of the increasing pressure.

Proposals

a) Make the CAP more responsive to changes in world prices.

b) Extend preferences to more low-income exporting countries, disadvantaged by existing CAP protection.

c) Reduce size of EC sugar qouta.

d) Compensate Mediterranean countries for enlargement of the EC.

e) Make one-off arrangements to cushion effect of higher food prices on nutrition levels in any countries adversely affected by reducing CAP protection.

See also: *The Common Agricultural Policy and the Less Developed Countries*, by Alan Matthews, Gill & Macmillan, Dublin, 1985

* * *

Sir Henry Plumb MEP

Chairman, European Democratic Group in the European Parliament

"Africa is capable of feeding its own people, and the agricultural industry of the UK and the rest of Europe can make a contribution if it accepts a modest transfer of resources away from the support of our own farmers towards aid for the impoverished and long-neglected peasant farmers of Africa. Britain could take the lead in proposing suitable changes. I would like to think, with my long experience in the British agricultural industry, that we had both the generosity and the foresight to see that this is in the best interests of both the Community and Africa. . . ."

Proposals

a) Reduce dumping of EC agricultural products on the world market by bringing EC prices more into line with those of our competitors. Take care, in doing so, to share the burden of social measures required to protect EC farmers and consumers.

b) Tougher standards for cereals should be imposed — to ensure that sub-standard wheat (unsuitable for food aid) would not be taken into intervention — thus encouraging farmers to grow other crops (e.g. rape-seed) for which there is a market.

c) The sugar Quota should be abolished over a 5-year period — thus reducing the amount of sugar grown in the EC and benefiting sugar cane growers in the Third World.

d) More should be spent on structural measures to encourage farmers to take land out of production and turn to forestry, or, on a temporary basis, set land aside as fallow.

e) Control of food aid should be removed from DG VI (Agriculture) and given to DG VIII (Development).

f) ACP countries should be given the chance to discuss their concerns over changes in the CAP — and EC countries should listen to them.

g) There should be an EC agricultural training programme in the Third World as a way of transferring resources.

Brian Rutherford
United Kingdom Agricultural Supply Trade Association Ltd (UKASTA)

1 Animal feed is produced by least cost formulation techniques on the computer. The technique arrives at the cheapest formulation practical with known ingredients to the specification desired for each class of stock. The animal feed industry's job (as it sees it) is to produce the most economic (cheapest) ration with the available raw materials.

2 Since 1973 the EC has had self-sufficiency as a target for protein. The Commission has been putting pressure on the animal feed industry to use more cereals.

3 Price reductions for cereals are unlikely to lower production; farmers will simply use more fertilisers, pesticides, etc, in order to maintain or even increase production.

4 Over the past 10 years, the Commission has put enormous pressure on manufacturers to use more cereals to reduce the surplus. But the answer to the cereals surplus is not to put more constraints on animal feed manufacturers to use more cereals, but rather to contain or reduce the price of cereals to make them more competitive.

5 Major imports are cassava and protein — but local EC production of rape-seed and sun flower seed is rising — so imports of protein are likely to fall.

6 But self-sufficiency is an expensive pastime; the EC's subsidy on rapeseed is currently £150/tonne.

7 Farmers like imported raw materials because they reduce the price of animal feeds. Because of the high prices of grain in the EC, we are able to pay high prices for imports.

8 But farmers are demanding less variability in animal feeds — which works against Third World suppliers, who find it difficult to maintain quality.

9 Many of the imports are by-products, and cannot yet be productively used in the Third World, so it is best to export them, even at low prices. So Third World exports of these by-products is likely to continue.

10 The lower the EC price of cereals, the lower the price will be for the by-products, both home-produced and imported. But probably the price of cereals will remain high enough to induce Third World countries to export to the EC.

* * *

Paul Rynsard
The Green Party, Agriculture Working Group

1 The CAP's protectionism increases pressure on Third World countries to export agricultural cash crops.

2 Lower world prices mean that Third World countries must export more to repay their debt.

3 These pressures work against 'sustainable agriculture' and encourage over-exploitation of the soil, to the detriment of many Third World people today — and many more in the future.

4 'Sustainable agriculture' uses science-based appropriate technology to take into account long-term ecological factors as well as short-term economic ones. It provides a way forward, both in the EC and in the Third World.

* * *

Dr David Seddon
Overseas Development Group, University of East Anglia

Food riots in North Africa

1 The USA and the EC countries, particularly France, are fighting a trade war over grain exports to Morocco and Tunisia.

2 Competing credits from France and the USA for cereal imports enable the governments of Morocco and Tunisia to continue to neglect their own agriculture — to the detriment of the poor.

3 Both countries follow an 'open door' strategy of development — putting resources into industrial development and replacing agricultural output with cheap imports from abroad, which conveniently also keep down urban wages.

4 Cheap grain enables these two governments at least to continue with a policy that hurts the poor.

5 Cuts in subsidies and drastic price rises for food grains in 1983 (at the instigation of the IMF) led to serious riots in early 1984. They were severely repressed. Those most hurt were the poor.

* * *

F. E. Shields
National Federation of Young Farmers' Clubs

1 The CAP can go one of two ways:
— more protection but how do we control surpluses
— recognise a world responsibility and leave a section of its market to the Third World

Either way Quotas will be the answer. They will provide Third World farmers with a pre-determined section of the market, and give EC farmers set targets for production.

Huub Venne

Free University of Amsterdam

1 The cassava industry in Thailand started from the coincidence of large numbers of unsettled people from the war zones adjoining NE Thailand and the possibility of the EC market for cereal substitutes. If these displaced people had not grown cassava, they would have grown something else.

2 Also, there were political and military motives for developing the transport infrastructure.

3 The EC's aid programme can be seen as a kind of compensation — and thus a precedent. But there were also strong military and political reasons for doing something to avoid dissatisfaction in the area, thus leaving the way open for communism.

4 The quota (or Voluntary Restraint Agreement) costs Thailand £100m/yr — but the compensatory aid is only £32m over 5 yrs.

5 If we want Third World countries to have more money and to be less dependent on exports, we should encourage them to do more local processing.

6 It would not help the situation in Thailand if the EC stopped buying cassava — that would only increase poverty. The answer is gradual diversification — but there is no clear profitable alternative.

* * *

Nick Viney

Dairy and Sheep farmer, Dorset

1 Existing surpluses should be reduced to a strategic level as fast as possible — by quotas (milk) and set-aside (cereals). Prices must be held down only for consumers, but also to ensure that farmers don't lose the market to synthetic products.

2 To prevent the USA from getting any markets vacated by the EC, control of agricultural production should only be done in agreement with the USA. The Panel should present its Report in the USA with a view to achieving this.

3 It should be accepted that every country in the world has the right to produce 80% of its own food if it wants to.

4 Third World countries should therefore have the right to erect tariff barriers, without which their agriculture is unlikely to develop.

5 Any country which feels its right to produce its own food has been contravened should have the right of appeal to some sort of legal court (e.g. UN).

* * *